3 200203 524

GW01338925

WITNEY LIBRARY
WELCH WAY
WITNEY OX8 7HH
TEL. WITNEY 70386

**OXFORDSHIRE
COUNTY COUNCIL
Cultural Services**

**FOR USE
IN
THE LIBRARY
ONLY**

The Flora of DITCHLEY

Wild Flowers of an
Oxfordshire Estate

The Flora of DITCHLEY

Wild Flowers of an Oxfordshire Estate

A n i t a   J o   D u n n

With photographs by the author

Estate Map by
Simon Dorrell

Drawing of Wild Service-tree by
Ann Buckmaster

Copyright © Anita Jo Dunn 1993

ISBN 0–9521310–0–5

Published by Dr. Catherine Wills
Sandford St. Martin, Oxfordshire.

Produced by Alan Sutton Publishing Ltd.,
Stroud, Glos.
Printed in Great Britain

DEDICATED

to the memory of

MARTIN WILLS

(4 October 1952 – 16 April 1992)

Contents

Acknowledgements	ix
Foreword	xi
Introduction and Map	1
Areas of the Ditchley estate covered by the survey	4
The Ditchley landscape and its history	5
Old woodlands and new plantations	10
Ancient woodland plants and the link with Wychwood Forest	16
Records of a century ago	19
The Wild Service-trees	22
Woodland grassland and old pasture	26
Cornfield flowers	29
Butterflies and plants	32
References	34
List of plant species	35
Colour plate section (between pages 28 and 29)	

Acknowledgements

I am greatly indebted to the late Mr. Martin Wills, the owner of Ditchley, first of all for his kindness in allowing me the freedom of the estate for the purpose of the botanical survey, a privilege I have valued and enjoyed for over seven years; for the keen interest he expressed, unfailingly, in the progress of the project; for valuable information about the estate, gained sometimes through our correspondence, but especially during a few memorable conversations; and finally for his generosity in setting aside a fund, administered by his sister, Dr. Catherine Wills, for the publication of this book. His belief in the importance of the survey was a great encouragement, and without his warm and enthusiastic support it would not have been possible to bring the results to fruition in this way. For all these kindnesses I remain deeply grateful.

I wish to thank his parents, Sir David and Lady Wills, for their kind hospitality, and Catherine for timely assistance and advice. My gratitude is due no less to Martin's cousin, Lady Rosemary FitzGerald, a friend and professional botanist, who with Martin had initiated the survey, and invited me to take over the recording work when she was no longer in a position to continue it. From the beginning Rosemary has shared my delight in many of the discoveries, and during the present winter found time not only to read the manuscript but also to contribute a number of valuable comments and suggestions. I have welcomed and enjoyed her involvement.

I am grateful to Mr. Peter Cooper, Land Agent, Laws and Fiennes, for copies of estate maps, information about the woodlands, and the loan of records relating to them.

I also wish to thank the following members of the estate staff: Mr. Harold Wakefield, head forester until his retirement in December 1992, whose long and close association with Ditchley and its woodlands has been the means of greatly enhancing my knowledge of both; Mr. Chris Eaton, gamekeeper, whose discovery of Herb Paris was the highlight of 1989, and whose frequent observations on many of the plants and their locations have added important information to the records; Mr. Stan Downes, farm foreman, Ditchley, and Mr. David Parsons, farm foreman, Fulwell, for providing details of field names, boundaries, crops and other farming matters; and Mr. David Burden, farm foreman, Spelsbury, for similar assistance and for his help in tracking down the Great Horsetail. During many conversations all have freely shared their knowledge of the estate with me and patiently answered my many questions.

I am grateful also to Mr. Peter Sheasby for a copy of the Wychwood plant species' list; Mr. John Campbell, County Records Officer, Biological Records Centre, Woodstock, for Ditchley butterfly records; and Miss Serena Marner, Manager of the Fielding-Druce Herbarium, Department of Plant Sciences, University of Oxford, for expert help with herbarium specimens and literature.

Permission to include quotations from the works or correspondence of Mr. Frank Emery, Mr. Alan Mitchell, Dr. Oliver Rackham, Mr. Patrick Roper and Miss Beryl Schumer has been granted by the authors or their publishers.

Jo Dunn
Charlbury, 30 December 1992.

Foreword

'Love of the land and care for its heritage' is often claimed as one of the finer characteristics of the English, but in the view of botanists and conservationists the second half of the twentieth century seems to have changed that into 'Exploitation of the land and greed for its products and profits'. Large lowland farms which still have diversity of land use, where high productivity has not been forced at an irrecoverable cost to the countryside, are becoming frighteningly rare, but Ditchley is one. The land here still has living plant communities which show its history, still has a mosaic of field and hedge, wood and water where wildlife can co-exist with economic agriculture, and this book celebrates its history and survival.

Two people have been essential to this celebration. My late cousin Martin Wills, most recent owner of Ditchley, loved this land passionately, in particular its trees and woodland. His intelligence, individuality, humour and curiosity, his respect for the past combining with interest in the present and care for the future, made his management of the estate both creative and responsible. To the very end of his long and courageously borne illness he felt unflagging fascination for Ditchley matters, taking particular delight in the progress of this book whose author, Jo Dunn, is the second essential character.

Providence was helpful when the 'Ditchley Plant List' was thought of, Jo's home being almost on the estate and her importance in Oxfordshire botany already established; but during the seven years of recording it was her own vision and enthusiasm, with her skill in the field and meticulous historical research, which expanded the original modest concept into this delightful and scholarly local flora. Landowner and field botanist too seldom feel as one, but this book is a fine memorial to a landowner who believed that nature and land management were inseparable, and it gives hope and encouragement to those who believe that the countryside deserves a future. I am honoured to be able to write these few words, with respect and affection, for the two rare people who created it.

Rosemary FitzGerald
Lilstock, 14 January 1993

INTRODUCTION

At a time when Britain's wild flowers are under constant threat from a growing number and variety of pressures upon their habitats, some private estates have become important refuges for many plants which are becoming rare in the surrounding countryside. Ditchley, in west Oxfordshire, is an outstanding example of one such estate. Its flora, in contrast to that of the fields, hedges and verges beyond its gates, is still remarkably rich, reflecting its physical, ecological and human history in a part of Oxfordshire rich in traces of the past. Its lands, which lie on the Great Oolitic Limestone, have sections of Grim's Ditch, an Iron Age earthwork, within its borders; evidence of long Roman occupation; and a documented history showing that Ditchley was once part of Wychwood, one of the great hunting Forests of Saxon, Norman and medieval kings.

Although today the estate's 4,646 acres are devoted primarily to farming and forestry, they retain fascinating inheritances from the past, including a wealth of different plant habitats: parkland and pleasure grounds; remnant deciduous woodland; plantations on ancient woodland sites; clearings, rides and tracks; lake, streams and woodland ponds; many hedges; and old pasture and arable fields.

Though it was known that over the past century a few botanists, including the renowned George Claridge Druce, occasionally recorded plants in limited areas of the estate, it had been clear for some time that the flora of Ditchley merited a long-term comprehensive survey. Once the decision was made to do this, the field work began in earnest in 1985 and continued for seven years. More than 400 wild flowers, grasses, sedges, rushes, ferns, shrubs and trees were recorded, including several species now very scarce in Oxfordshire. At the same time, because it was recognised that the plant communities were long established, and partly of ancient woodland origin, it was decided also to explore Ditchley's historical ecology, mainly through records held by the Bodleian Library and the County Archives Office. This proved to be as fascinating as the plants themselves, and as these two aspects of the survey – the botanical recording work and the historical research – complement each other and are inseparably related, the results of both studies are included in the *Flora*.

It should also be explained that although precise locations are not given in the plant list, each species has been separately card-indexed to show, among other details, where and when it was recorded, so that botanists who may undertake future surveys at Ditchley will have a data-base from which to work.

The results of the survey have amply justified the belief in Ditchley's importance as a 'reservoir' and sanctuary for a host of uncommon and locally scarce wild plants. At the same time, the survey has added 240 new records to those already held for Ditchley by the vice-county Recorder for the Botanical Society of the British Isles, and has made a contribution to the new *Flora of Oxfordshire*.

The Ditchley Estate

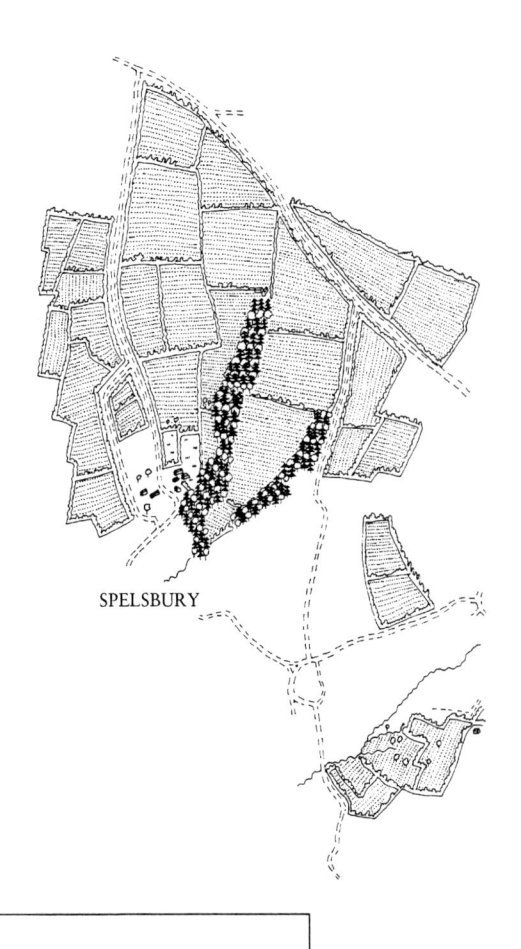

SPELSBURY

KEY:

Pleasure Grounds

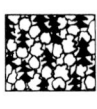

Parkland

Arable Land

Long Ley

Ancient Woodland Site, linked with Wychwood Forest; now planted with beech-with-conifer

Beech-with-Conifer Plantation

(Land use as in October 1992)

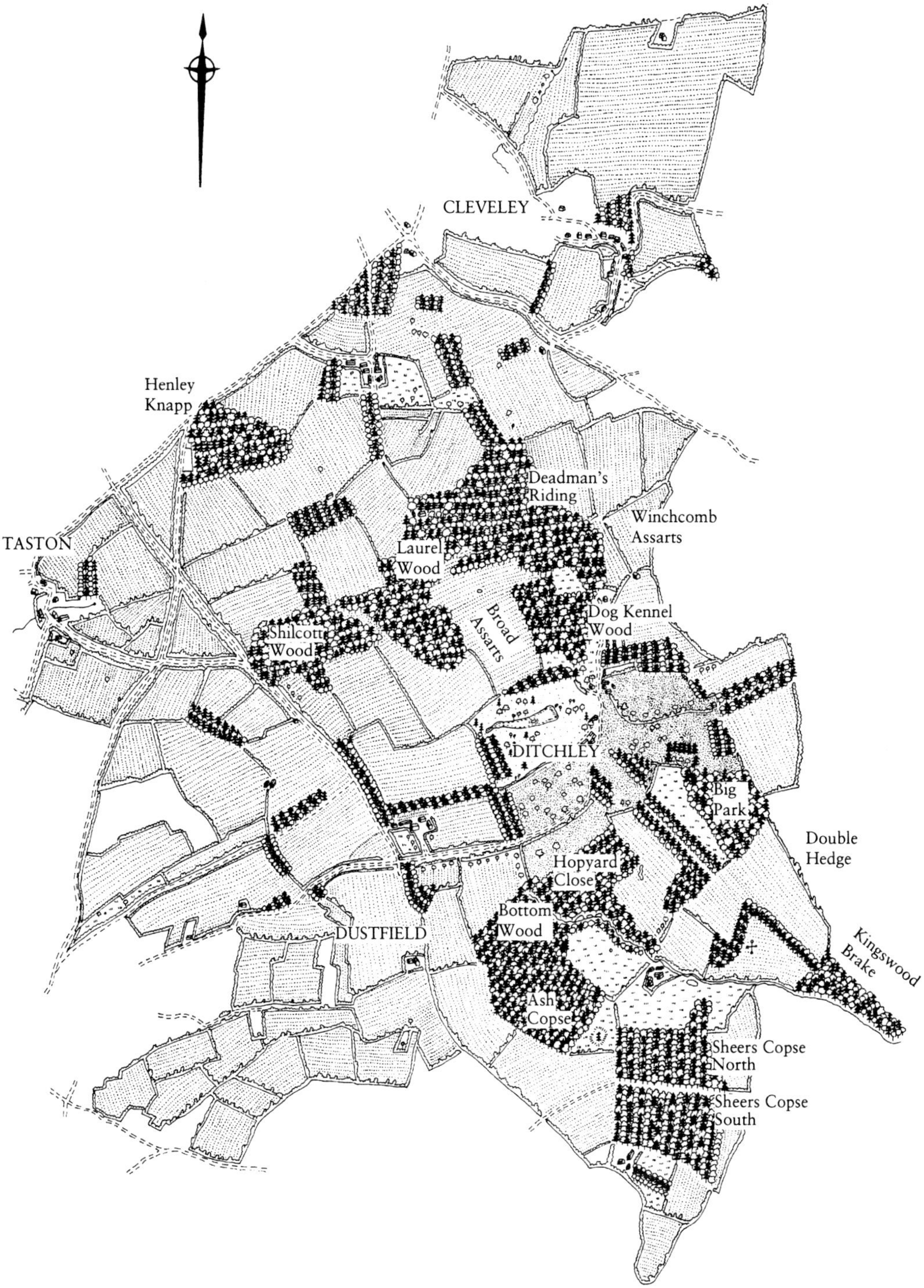

CLEVELEY
Henley Knapp
TASTON
Deadman's Riding
Winchcomb Assarts
Laurel Wood
Broad Assarts
Dog Kennel Wood
Shilcott Wood
DITCHLEY
Big Park
Double Hedge
Hopyard Close
Bottom Wood
DUSTFIELD
Kingswood Brake
Ash Copse
Sheers Copse North
Sheers Copse South

AREAS OF THE ESTATE COVERED BY THE SURVEY

Ditchley

1 km Grid Ref. No.

Ash Copse (including the compartment formerly known as High Wood)	SP 3819
Barrow Clump	3919
Big Park (formerly known as Box Wood)	3920
Bottom Wood	3820
Clarke's Bottom/Pintle Barn	3620/3720
Cleveley Bank (part of)	3923
Crooks Corner	3822
Deadman's Riding	3822
Devil's Pool (below Lodge Farm), and stream to Kingswood Bottom	3919
Devil's Pool Plantation	3920
Dog Kennel Wood	3821/3822
Double Hedge on Kiddington boundary	4020
Dudgley Pool	3822
Ellen's Lodge	3822
Fulwell (East)	3723/3823
Fulwell (West)	3722/3723
Henel Plantation	3722
Henley Knapp	3622/3722
Hopyard Bank and Hopyard Close	3820
Hundley Way	3620/3720
Jollys Ricks bank	3823
Kiddington Lodge	3921
Kingswood Bottom	4019
Kingswood Brake	4019
Lake, including surrounding grassland and woodland	3821
Laurel Wood	3822
Little Park	3820/3821
Lodge Farm and environs	3919
Model Brake (North)	3720/3721
Model Brake (South)	3820
Model Farm and environs	3820
Newbarn Farm pastures and perimeter of 'The Bog'	3918
Norman's Grove	3721
Old Grubbs	3721
Pleasure Grounds	3821
Sheers Copse (North)	3919
Sheers Copse (South)	3918/3919
Shilcott Wood	3721
Sladhollow, Cleveley	3923
South Front Plantation (East)	3921
South Front Plantation (West)	3920/3921
Spider's Spinney	3720
Starveacre	3920
Timberyard	3921

Spelsburydown Farm and Lower Farm, Taston

Blaythorn (West of Dean)	3322
Coate (adjoining Taston Brook)	3521
Lower Farm meadow	3621
Spelsburydown Farm, including stream and adjacent plantations	3422/3423

(The above areas include adjoining cultivated fields)

THE DITCHLEY LANDSCAPE AND ITS HISTORY

PHYSICAL FEATURES

Ditchley is situated in west Oxfordshire at an elevation of about 150 metres above mean sea level. It lies within the parishes of Charlbury, Enstone and Spelsbury, between the rivers Evenlode and Glyme, tributaries of the upper Thames. In addition to 3,665 acres of land at the heart of the estate and at Cleveley, there are 981 acres at Spelsbury and Taston. These 4,646 acres include approximately 170 acres of parkland and pleasure grounds, 850 acres of plantations, 160 acres of pasture, over 3,000 acres of arable land, and several acres of farm buildings, cottages, gardens, roads and quarries, etc.

In this part of the county the underlying rock is Great Oolitic Limestone (Jurassic), which contains numerous bands of associated sedimentary rocks and clays, including those known to geologists as Chipping Norton Limestone, Sharp's Hill Beds and Lower Lias Clays. There are no sands, gravels, or glacial deposits. An interesting well sinking made on the estate, near Charlbury Lodge, in 1894 is shown to pass first through 4 ft. of humus and marly rubble, then 10 ft. of 'shelly limestone'(building stone). Below this were thin bands of 'soft marly limestone', 'hard blue Marl', 'black Clay with shells' and, lower down, 11 ft. of 'hard blue Marl and Clays'. The subsequent water supply apparently proved to be poor and insufficient, the main reason for this being the inability of the clay beds to make a good water line. Nevertheless, there is still an unfailing supply from the old source at Spurnell's Well. A few springs emerge in hollows or below banks but disappear in dry periods; one feeds a stream that in the wettest months of the year flows south-eastwards into Kingswood Brake, but is dry in summer. Three streams flow through or border the Spelsbury and Taston lands.

The soils are brashy and, on the whole, well-drained, except where there are clay bands near the surface. The average annual rainfall is about 642 millimetres (25 ins.).

Viewed from the parkland, the landscape is in part gently undulating, in part level; while in other places, often completely hidden from view, are steeper slopes and banks, ditches and hollows. It is this variation in the estate's topography and the pattern of land use which provide such a rich diversity of plant habitats. In addition to the pleasure grounds and fine parkland with their mature and young beech, lime, oak, sycamore, horse chestnut and occasional field maple trees, there is a 4^1/$_2$-acre lake; the streams and springs mentioned above; several woodland ponds; relict deciduous woodland and new plantations; wide rides, lights (grassy compartments) and tracks; old pasture and arable fields; and many hedges which Martin instructed should be left to grow to 7 ft. before being trimmed, for the benefit of wildlife. Also, by happy coincidence, Ditchley is in limestone country, and it is the calcareous soils of limestone and chalk which, by their very nature, support the richest variety of plant species, including some of our rarest wild flowers. Acidic soils are much poorer.

FROM PREHISTORY ONWARDS

It has long been established that Ditchley is set in countryside anciently settled by man. Large-scale maps show where prehistoric flint implements have been found on its lands. Parts of Grim's Ditch, a circular Iron Age earthwork which once enclosed 22 square miles of country, run through it. There are two small earthworks (barrows), and on its northern boundary stands an ancient Hoar Stone. Maps also mark the site of a Roman villa, one of several in the district, that stood little more than half-a-mile from the present mansion, near the estate's south-eastern border. Excavated in the last century, and now visible only from the air as crop marks, it dates from A.D.70. The 16-room stone building, which replaced a timber farmhouse, was surrounded by a wall. This enclosure also housed several out-buildings, including a well, a threshing floor, and a granary said to have been capable of storing corn from 1,000 acres of land; and beyond the borders of the farmland would have been a great tract of largely unexplored woodland. It remained what was clearly a prosperous settlement until the villa was partly burnt down in *c.* A.D.200. It was then forsaken for about one hundred years, rebuilt in the 4th century and finally abandoned in the 5th century when the Romans left Britain.

It was the Neolithic settlers, however, who arrived some four thousand years before the Romans, who were the first agriculturalists. It was they, with their primitive tools, who first laid hands upon the primary woodland or 'wildwood' that clothed Britain after the retreat of the last Ice Age some 12,000 years ago. Active clearance of the trees that grew in the primeval forest (sallow, birch, pine, hazel, oak, ash, elm, alder and small-leaved lime) began in the Iron Age and continued apace throughout the Roman and Anglo-Saxon periods as the population grew and more land was needed for cultivation, until eventually there was more farmland than woodland. But while this was true of the country as a whole, and particularly in the river valleys, in the Wychwood area the picture was rather different.

In her paper *The Evolution of Wychwood to 1400: Pioneers, Frontiers and Forests (1984)*, Beryl Schumer says:

> Both archaeological and place-name evidence imply the presence of woodland in the Saxon period, and it seems that in the Wychwood region that period was not one of expansion, as is usually assumed, but of contraction of settlement, in that areas which had been occupied from the Bronze Age to the Roman period were apparently uninhabited and used only as wood-pasture.

This contraction of settlements in the Wychwood region continued, apparently, for centuries. This being so, it can reasonably be assumed that there followed an encroachment and establishment of secondary woodland on much of the land that had previously been cultivated; and that this woodland, which was perhaps not very different from the original primary woodland, survived through the Saxon and Norman periods to become part of the great royal hunting Forest of medieval Wychwood.

On a map of Domesday England, 1086 (redrawn by J.S. Garnons Williams in 1984), in the Bodleian Library, 'Spelesberie', 'Corneberie', 'Wodestoch', 'Cumbe', 'Stuntesfeld' [Stonesfield], 'Optone' [Wootton], 'Radeford' and 'Sandford' are shown as 'vills' in west 'Oxenefordscire'. Although Ditchley is not included, this is not to assume that it did not exist at that time (Domesday Book omitted many significant and minor details). Beryl Schumer's research has shown that some Wychwood hamlets were in existence before 1086, and that '. . . there were others still smaller, whose date of origin it is impossible to determine. They were situated at the boundary between two or more manors, within or at the edge of the woodland . . . One of these is Ditchley, a small hamlet at the meeting-place of the woods belonging to Spelsbury, Enstone and Bloxham. The earliest references suggest that it was then the home of the keeper of Bloxham Wood, with extensions of the hamlet into Enstone and Spelsbury occurring later.' The name Ditchley, spelt Ditcheleye in a record of 1208, means 'a clearing in the forest beside the ditch', a description which in itself implies that it must have originated long ago as a forest hamlet.

Also in the Bodleian Library are records which show that private ownership (by, it would appear, lesser nobles, lords of the manor or gentry) of the Ditchley lands dates back at least to the 13th century, and possibly beyond.

An expanding population in the 13th century increased the need for more land for cultivation, and in forest country this could only be achieved by assarting (a word derived from the Old French *essarter*, 'to grub up trees'). The medieval owners of Ditchley met this need by assarting parts of the surrounding woodland by grant from the king. Winchcomb Assarts, on the north-eastern border of the estate, was cleared in 1307. Broad Assarts, Common Assarts, Ditchley Assarts, Rough Assarts, Western Assarts and Wootton Assarts, which, with the exception of Broad Assarts, lie along the eastern perimeter of the estate, were probably also formed out of woodland during this period, as records show that 200 acres were assarted in Ditchley in 1345. In 1853 Winchcomb Assarts was described as 'Common Waste'.

In his chapter on 'The medieval expansion', in *The Oxfordshire Landscape* (1974) Frank Emery writes:

> . . . assarting continued through the thirteenth century before the covert of the forest (i.e. unbroken woodland with trees meeting overhead) was defined in 1279. By a new perambulation of 1300 only the royal woods were left inside the forest [of Wychwood], and the king followed a policy of granting wastes to those who wished to assart them. Even so, Wychwood covered some 50,000 acres from Woodstock to Taynton, and from Ditchley to Witney.

An account of this perambulation of 1300 describes how part of the Forest boundary proceeded:

... Thence, by the middle of Dustelsfeld, to Grimesdich; and thence, between the woods of Bloxham and Spelesbury. Then from Grimesdich, by the corner of the wood of Bloxham, near the wood of Spelesbury, to Dichelehegg; and straight from Dichelege, as the hedge extends between the lands of Henry de Dichelege and Agnes de Bloxham, to the before-named Grimesdich, following the boundaries between the wood of Bloxham and that of Ennestan, called Le Boxe.

The *hegg* in 'Dichelehegg' is probably a form of the Anglo-Saxon word *hege*, meaning hedge; as implied in the passage above, many hedges then, as now, represented boundaries. And the *feld* in 'Dustelsfeld', is not a 'field' but, to quote Dr. Oliver Rackham, author of *Ancient Woodland* (1980), '. . . an open space in sight of woodland with which to contrast it.' Dustfield, as it is now known, is on Ditchley's southern edge. Closely associated with it is a wide green lane known to be of pre-Roman date, and it is tempting to think that it may have been a woodland trackway or mereway of even greater antiquity.

Although encroachment upon the woodland continued, this had little effect upon the importance of the great royal Forest of Wychwood. Like all the king's Forests, Wychwood contained hamlets, farms and common land as well as woodland and was of great prestige, social and economic value to a succession of monarchs, not least because they, or their huntsmen, could pursue the royal deer conveniently from the lands they owned, or from their palaces. One palace was at Woodstock and is known to have been used by Henry II (1154–1189). The Forest was also of great importance to those of the king's subjects – the landowners and commoners – who owned or were granted rights and privileges within it, for example, the liberty to pasture sheep or cattle in woodland or on common land, to fatten pigs in autumn on acorns and beech mast (pannage), fell trees, cut underwood, collect dead wood (estover), and burn charcoal.

In their turn the sovereigns appointed many Forest officers. One of these, in the 16th century, was Sir Henry Lee, who purchased the manor of Ditchley from Thomas Gibbons in 1583. For the sum of £400 he acquired '. . . two messuages (houses with their grounds), two gardens, two orchards, 400 acres of (arable) land, 40 acres of meadow, 200 acres of pasture, 250 acres of wood and 300 acres of furze and heath.' He was a close adviser to Elizabeth I and Ranger of Woodstock. In 1603, in the last year of her reign, the queen granted him a licence to enclose a park for deer at Ditchley. This event is recorded, surprisingly, in the form of a small printed notice pasted on to a large-scale Ordnance Survey map of 1881 in the collection of Ditchley manuscripts, maps and papers at the County Archives Office in Oxford. It reads:

> November 1603
> A Signet Licence was granted to
> Sir Henry Lee
> to enclose as a
> Park for Deer and other Game
> certain woods and woodgrounds
> of his own in
> Charlbury, Spillesbury, Toston
> Fullwell and Ditchley
> For which a Fee of £3 was paid

James I, successor to Elizabeth I, hunted at Ditchley in 1608 and 1610. The heads of the deer he killed are preserved in the mansion, together with six brass tablets briefly commemorating, in verse, the events of the chase. The names of two of the woods through which he and his men hunted, Deadman's Riding and Henley Knapp, are still in use on the estate today and appear on contemporary maps. 'Foxhole Coppice', from which a deer was roused on 24 August 1608, appears as 'Foxhole Wood' on an estate map of 1726 and later became part of Shilcott Wood. Another of the king's deer ran to 'Gorrel Gate', which is assumed to be the 'Rorill Gate' of 1881; it is close to the 'Rorill Wood' of 1703, now known as Laurel Wood, 'Laurel' perhaps being a corruption of the older name.

Sir Henry Lee's deer park remained enclosed for nearly 350 years. The house he built at Ditchley was replaced in the 1720s by the present mansion, built by one of his descendants, George Henry Lee, second Earl of Litchfield. The head forester, who retired in December 1992 after serving the estate for fifty-one years, forty-five of them as a forester, recalls that until 1951 or 1952, when the deer fencing was allowed to fall into disrepair, it was the duty of the maintenance or forestry staff to check the fencing every Saturday morning, to make sure that all the oak pales were in place, and to mend any gaps.

Although some replanting of trees had taken place between World Wars I and II, this followed the traditional practice in north Oxfordshire of growing Larch for fencing and Ash for firewood. After World War II the woodlands were found to be in a neglected condition and unproductive, though still important for the shooting.

Sir David Wills purchased the estate in 1953 and gradually introduced new management policies, including the restructure of the woodlands and modernisation of the farms.

Although for at least two thousand years man has left his imprint upon this piece of country, and during that time has modified its landscape, Ditchley retains its sense of history, its dignity and beauty. One of the finest estates in Oxfordshire, it is most fortunate in its natural inheritance and also in the circumstance that for centuries it has remained, and still remains, in private and often distinguished ownership.

OLD WOODLANDS AND NEW PLANTATIONS

THE OLD WOODLANDS

Once Ditchley's ancient association with Wychwood Forest had been firmly established through historical evidence, curiosity about the estate's old woodlands led to a rewarding search among documents in the County Archives Office. From a collection of woodman's accounts, timber valuers' estimates and reports, articles of agreement, memoranda, notices and maps dating back to the 17th century there emerged a fascinating picture of the earlier woodlands and the way in which they were managed. It was clear that although of course different in their composition, and serving different needs, they were no less important economically two or three centuries ago than the new plantations are today. Moreover, they abounded with human activity, were jealously guarded by their owners and protected by strict laws against damage and theft.

In addition to the value of the timber trees (predominantly oak), there was an important secondary income from the sale of underwood – the hazel coppice poles or, less often, poles from certain trees which were pollarded. Before 1853 the Ditchley underwood was cut every thirteen years, but after that date every ten years. This was done on a planned rotation through the different woods, an average of '39 customary or wood acres' being cut every year.

The oldest document among the papers examined was written in copperplate on parchment. It lists the 'Underwood sould at Lees rest, Anno Domini 1683' to 53 men from Charlbury, 63 from surrounding villages, 13 from Woodstock and three from Oxford. The lots were divided into 'lands', 'small acres' and 'great acres', these costing on average, respectively, about 12s.0d., £1.5.0d. and £6.6.0d. Each man would have been responsible for cutting and carrying or carting away the material from his particular lot. The underwood cut by these men, or by woodmen employed on the estate, was sorted for a variety of uses. There was a heavy demand for 'faggotts' (20,000 in one sale in 1853), for kindling, heating bread ovens, and as foundations for corn ricks; 'bonds' or 'bandes' for binding the faggots; stakes for sheep hurdles and hedging; 'hetherings' or 'binders' for securing the hedges as they were laid; 'sprays' (thatching pegs); and bean-poles and pea-sticks, some of which went to the estate's gardeners. There was also 'legwood', a term the head forester says is still in use today for bundles or pieces of faggot suitable for firewood. Lees Rest, the wood mentioned above, was bought by Sir Henry Lee in 1591 and, sadly, grubbed up in 1845–50.

Descriptions of the woodmen's tasks are no less interesting. 'A Bill of Disbursments in Ditchley Woodes from Michaelmas 1746 to Michaelmas 1747' accounted for, among other items:

'. . . Macking one thousand 6 hun^d. and 70 faggots at 1s.6d per hun. in the parck, £1.4.10½d'
'. . . 5 Dayes Work for Cutten of faggots bandes, £0.5.0d'
'. . . making of 17 poles [measurement] of hedgen and Ditchen at the three brackes ['brake' was probably a general term for a thicket of bracken, thorn or gorse].'
'. . . maken of 40 poles of Quicksetting [Hawthorn] and a Double hedge at Winchcom Sattes [Assarts] at one Shillien and 4 pence pr pole, £2.13.4d'
'. . . 5 Dayes Work a Lapping [lopping] for the Deear at 1 Shillin pr day, £0.5.0d'
'. . . 19 Dayes Work a Stoping of Gapes, £0.19.0d'

The bill ends with the words: 'My Yeares Wages Due att Michaelmass Last, £10.0.0d'.

Also in 1746, several local people, 'thomas Ceates' among them, were sold 'penney Thornen faggots' from 'Shears Copice'. In the following year, '3 score Oaken lap faggots from Dearnel Wood [now Dogkennel Wood]' were bought by 'heneary haines and Widdow Yoxson', and another sale was to 'Mr.pricket, one pasell of Lap And too fiearwood trees'.

A similar 'Bill of Disburstments' for 1773/74 records that one shilling a day was paid for the following work: 'Cutting and Loading of Bushes . . . Hedging the Turnips [presumably to protect them from rabbits, of which there were vast numbers, and livestock] . . . Grubbing the Elms and faggotting the Wood . . . Cutting and Ditching the Brake . . . Quicksetting and hedging . . . Cleaving of Wood . . . Lapping and underhighting the Trees in the park'. 'Cutting out 3 loads of hardwood' was paid at the rate of one shilling a load, while 'Polarding and making 9 hun^d and 50 faggotts' fetched two shillings per hundred. Just over one hundred years later, in the 'Woodman's Accounts, 1897–1915', the tasks also included: 'cutting and barking [Oak] trees . . . trimming rides . . . limbing oak trees in Ash Copse . . . cutting and lapping 7 Spruce . . . felling trees . . . weighing bark Chipping Norton machine [weighbridge]', as well as the customary coppice work.

An early document of special interest is 'A Memorand of Tho. Cras about the brake at Kingswood Bottom'. Part of this reads:

> Dec 12 1710 Thomas Cras afermes before my Lord Litchfield [the First Earl of Litchfield, 1663–1718] that he did meet Thomas ffreeman Ditchley woodward at the brake . . . into Kingswood bottom where the old roade is to receive instructions from the said Thomas ffreeman for the hedging in and for cutting the underwood in the said brake . . . And the said Thomas Cras afermes that at the same time of the cutting the said brake that the said Thomas ffreeman marked up two oke trees with a hamer marke with the letters H.L. & the said Thomas Cras saw the two Oke trees cutt downe and barked . . . and the said Thomas Cras saw Stephen Collier fetch the two oke trees for the repaireing of Coldron Mill [Spelsbury].

Coldron Mill, recorded in Domesday Book, still stands surrounded by its millstreams, though it no longer operates as a mill. It was the subject of a grant by the Earl of Warwick in 1268; and in 1583, the year in which Sir Henry Lee bought the manor of Ditchley, it was noted that '. . . Margaret Damery, the miller's widow, still held Coldron Mill and half a yardland [about 15 acres] . . .' The two oaks felled for the repair of the mill would by necessity have been substantial trees, germinating from oaks that stood in Kingswood Brake some five hundred years ago.

The importance and value of the woodlands is implied in the following terse late 18th century notice which must have been circulated or displayed to warn those who thieved, or were tempted to thieve, from the Ditchley woods:

NOTICE TO PUNISH WOOD STEALERS

Whereas great damage has been done of late years to the Woods and Fences about Ditchley, by Persons unlawfully cutting the Underwood and live Hedges, and Stealing Posts Rails and other Fences set to preserve them, to the great destruction of the said Woods Plantations and Mounds; this is therefore to give Notice that Whoever shall be found Guilty of any or either of the said Offences hereafter, will be Prosecuted as the Law in such case directs – which by a late Act of Parliam^t is very severe upon such Offences.

The Act referred to would be the one brought into force on 24th June 1766 by George III which decreed that fines ranging from five to twenty pounds (or banishment to a 'House of Correction', with the prospect of being 'thrice Whipt.') would be imposed on '. . . every Person who shall wilfully cut, break down, bark, burn, pluck up, lop, top, crop or otherwise deface, damage, or destroy, or carry away any Timber, or Trees likely to become Timber . . .' The trees specified were Oak, Beech, Chestnut, Walnut, Ash, Elm, Cedar, Fir, Asp (Aspen), Lime, Sycamore, Poplar, Alder and Birch. Also included were 'Underwood, Poles, Sticks of Wood, green Stubs [stumps]', and even roots, shrubs and plants.

Guardianship of the Ditchley woods was not only the responsibility of the gamekeepers. An entry in the 'Woodman's Accounts' for December 30th 1899 reads: 'W. Harrison – nightwork watching holly trees (5 nights) 11s.0d'. This must have been to prevent thefts of boughs before Christmas. Another entry, for October 1901 refers to a payment of 17s.6d to the same woodman for 'watching woods in nutting season'. These entries are repeated in several subsequent years.

Records for the timber trees show that there must have been a great quantity of oak in the woodlands. One sale of timber on 15th March 1853 included 1990 Oak and 740 Elm, Ash, Beech, Sycamore and Larch. Other sales included Chestnut, Hornbeam, Field Maple and Wild Service.

From these and other documents studied, including one of 1714 entitled 'The Measur [sic] of Ditchley Woods', which lists each wood by name and its size in acres, roods and perches, it was possible to identify accurately and then physically locate the old woodland sites. This was considered essential if a record was to be made of the ancient woodland plants that might have survived.

These brief glimpses of the old woodlands and the human activities associated with them echo Dr. Rackham's statement in *The History of the Countryside* (1986) that '. . . woods are indeed at the heart of historical ecology. They are inherently stable and long-lasting, and have outlived many changes in human affairs. They take us straight back to the Middle Ages, and . . . conjure up before us the wholly natural landscape into which civilization was first introduced. They contain in themselves evidence of at least a thousand years of care and use.'

THE NEW PLANTATIONS

At the time the estate was purchased by Sir David Wills in 1953 the woodlands were described as 'semi-derelict and unproductive', the more important trees – the Oak, Beech and Ash – being over-mature or in process of dying back. To restructure them a long-term plan was embarked upon with the object of introducing Beech, which grows well on limestone brash, as the main crop. At the same time, for both utility and variety, other trees, mainly native species, were planted with the Beech or in separate groupings: Wild Cherry, English Elm, Wych Elm, Holly, Field Maple, Norway Maple, Oak, Wild Service, Sycamore and Yew; and canker-resistant hybrid Poplars were selected for the wet, low-lying lands. Lime, Horse Chestnut and small numbers of Deodar (*Cedrus deodara*), Grand Fir (*Abies grandis*) and Coast Redwood (*Sequoia sempervirens*) and a few other non-native species were added principally for amenity value, as were many rare and exotic shrubs and trees for the enhancement of the Pleasure Grounds. Conifers chosen as nurse trees for the Beech were Corsican Pine, Douglas Fir, European Larch, Lawson Cypress, Norway Spruce, Scots Pine and Western Red Cedar (*Thuya plicata*).

Except for a few acres of relict deciduous woodland which were left for the benefit of wildlife, all the old woodland sites were replanted: Ash Copse (including High Wood), Deadman's Riding, Dog Kennel Wood, Henley Knapp, Hopyard Close, Kingswood Brake and Kingswood Bottom, Laurel Wood, Sheers Copse (North and South) and Shilcott. These cover 525 acres. In addition, about 325 acres of unproductive arable and other farmland, rough pasture, scrub and parkland were afforested. During the planting, which continued until 1991, saplings of naturally regenerating Ash, Elm, Oak, Sycamore and Wild Service were left to grow on. It is estimated that the final composition of the plantations will be approximately seventy-five per cent Beech, fifteen per cent Oak and ten per cent other hardwoods.

THE EFFECT OF CONIFER EXTRACTION ON PLANT LIFE

Over the past thirty-five years or so, during the planting up of the old woodlands and creation of new ones, it has been the practice to set three rows of nurse conifers, then three of beech, or other hardwoods, in repeated sequence. Selective thinning of softwoods – the spruce, fir, pine, cypress and larch – commenced in the 1970s, when the trees were about fifteen to twenty years old, and has continued throughout each subsequent winter. Because these operations cause disturbance of the soil and bring varying degrees of light and warmth into the plantations, and sometimes result in the widening of rides, the consequences for plant life are very beneficial and often dramatic. Long buried seeds are at last able to burst into life, giving rise in places to a luxuriant and colourful flowering of woodland plants.

Reports of these resurgences of plant life often came from the gamekeeper or head forester, whose duties take them into the woodlands every working day. In the spring of 1987, for instance, word came that Columbines were flowering in a place where they had never been seen before. This was in a 1962 plantation where thinning of the Norway Spruce in 1986 was followed by the toppling of a few more in strong gales in March 1987. The plants had germinated in the open ground of the wind-gap.

In the spring of 1988, after the removal of several rows of Scots Pine from a 1964 plantation, dense patches of Yellow Archangel (*Lamiastrum galeobdolon*) and Bugle (*Ajuga reptans*) brightened the ground in several places. Then in mid-June, nearby, a mass of flowering Wood Vetch (*Vicia sylvatica*) was found sprawling over the dead brushwood, or, taking a grip on the rough bark of standing pines, climbing vertically to heights of up to ten feet (3 metres). It was a beautiful sight, and a splendid new record. In another plantation not far from this one, which is also on an ancient woodland site, heavy thinning and in places complete removal of the softwoods has dramatically increased the bluebell population. They are as dense now as they must have been in the past and are one of the glories of the estate. This was a place which Martin liked to visit every spring, and when the work of felling the conifers was in progress he asked that the trees should not be dragged out but cut into the required lengths on the ground and lifted out, to prevent unnecessary damage to the plants.

There was another fascinating example of plant regeneration in 1989, after three rows of Corsican Pine and three of Douglas Fir were removed from opposite edges of a 1971 plantation which is divided by a ride. The trees were dragged out and the brushwood burnt in several heaps along both sides of the ride which, as a result of the felling, had been widened to about fifty feet (15 metres). That summer, in the bare disturbed soil where the trees had been, and in the roughed-up ride, a host of different plants were recorded. The list included about 150 vegetative plants of Deadly Nightshade (*Atropa belladonna*), 83 Great Mullein (*Verbascum thapsus*), over 20 Common Gromwell (*Lithospermum officinale*), a few patches of Woodruff (*Galium odoratum*), several Weld (*Reseda luteola*), one Columbine (*Aquilegia vulgaris*), and a mass of different thistles, including rosettes of the handsome Woolly Thistle (*Cirsium eriophorum*). Even more astonishing were four plants of Thorn-apple (*Datura stramonium*), a native of central and southern America. Three of these had arisen, phoenix-like, from the ashes of the foresters' bonfires, and it was interesting to learn that in Virginia, USA this plant is known as Fireweed, from its habit of appearing after fires. For an annual, it can put on tremendous growth in one season. By the time the largest of the four plants had matured it stood just over four feet high and measured over five feet across (1.32 × 1.68 m). Although the Thorn-apple is sometimes grown in gardens, it is uncommon in the wild, occurring sporadically on rubbish tips, waste ground and cultivated land, usually in a hot summer, as was the case in 1989. How the seeds came to be in a woodland ride is a mystery. They are known to remain viable, under favourable conditions, for a very long time. In his book *Weeds and Aliens* (1961), Sir Edward Salisbury cites an example of Thorn-apple plants appearing in abundance after old ground in a demesne in Anglesey was broken up after having lain undisturbed for at least a century. The whole plant is highly poisonous. The large, pure white, trumpet-shaped flowers are sweetly scented and open in the evening to attract night-flying insects. The leaves have a somewhat malodorous smell and are still used in restricted measure by medical herbalists in the treatment of asthma.

These instances of the renewal of plant growth after conifer extraction from plantations on ancient woodland sites graphically illustrate the resilience of plants; and as this aspect of woodland management will continue for many years, there will be opportunities in the future to monitor other changes that will take place in the long transition from oak-over-hazel to predominantly beech woods.

Ancient Woodland Plants and the Link with Wychwood Forest

Encroachments upon the natural forest that long ago clothed this great tract of country in west Oxfordshire continued into the present century, so that today Ditchley and the fragment of Wychwood that survives are separated by about three miles of open country. For the historian, tangible evidence of their once close relationship is now often elusive and tenuous, or has been lost altogether; but for the botanist interested in comparing the floras of Ditchley and Wychwood, an opportunity is given to establish a link between the two which promises to be no less valid and unequivocal than the historical one but, at the same time, very much alive.

To prove a connection there were three closely related questions to answer: what evidence was there of the type of plant community associated with lowland oakwoods; how many, and which, ancient woodland plant species had persisted; and how did the flora of Ditchley compare with that of present-day Wychwood? The search for answers was concentrated on Ditchley's remnant decidous woodland, the margins of plantations and any open spaces within them, and adjoining rides and clearings. All these places are on the ancient woodland sites. It did not take long to settle the first query, because it was precisely in these areas that oakwood plants were found – plants that had undoubtedly descended from those which had flourished for centuries beneath the oaks, ash and hazels of the earlier woods. A long list of these species includes Wood Anemone (*Anemone nemorosa*), Wood Sedge (*Carex sylvatica*), Meadow Saffron (*Colchicum autumnale*), Enchanter's Nightshade (*Circaea lutetiana*), Dogwood (*Cornus sanguinea*), Spindle (*Euonymus europaeus*), Bluebell (*Hyacinthoides non-scripta*), Holly (*Ilex aquifolium*), Yellow Archangel (*Lamiastrum galeobdolon*), Honeysuckle (*Lonicera periclymenum*), Wood Melick (*Melica uniflora*), Wood Sorrel (*Oxalis acetosella*), Primrose (*Primula vulgaris*), Buckthorn (*Rhamnus catharticus*), Wych Elm (*Ulmus glabra*), Early Wood-violet (*Viola reichenbachiana*) and Guelder-rose (*Viburnum opulus*).

It was less easy to answer the question on ancient woodland plants. This is because botanists tend to disagree when discussing which individual species should be selected for inclusion in a list of plants considered to be indicators of ancient woodland in Britain. (The term 'ancient woodland' is used by Dr. Rackham to distinguish woods which have been in continuous existence since before 1700.) This is understakable because although there may be no doubt that a particular wood fits this definition, its flora is not necessarily typical of all such woods. During their evolution a variety of different influences have been brought to bear upon them, and therefore upon their plant communities. These include the geographical situation of the wood, its size, soil type, history, and management.

This being so, it was thought best to consult three authoritative sources of reference on the subject. The first was Dr. G.F. Peterken's paper 'A Method for Assessing Woodland Flora for Conservation Using Indicator Species' (*Biological Conservation*, Vol. 6, No. 4, October 1974). Another was *Woodland and Wildlife* (Whittet Books, 1992) by Keith Kirby. The third was a list entitled 'Woodland plants in Southern England which are "most strongly associated with ancient woodland and are typical components of botanically rich ancient woodland communities" '; this was prepared by Dr. Richard Hornby and Dr. Francis Rose and published in *Woodland Heritage* (1990) by Peter R. Marren.

As Dr. Peterken's list is based on sites in Central Lincolnshire, and Keith Kirby's on '. . . ancient woodland over the widest range', it was decided to use the third one, which is appropriate only to ancient woodlands in Hampshire, Wiltshire, Buckinghamshire, Berkshire and Oxfordshire. Of the 99 species on this list 49 (49%) occur at Ditchley. If Dr. Peterken's list had been used the count would have been 26 out of 49 (53%), and in the other case 31 out of 50 (62%).

The list drawn up by Dr. Hornby and Dr. Rose is sub-divided into Trees and Shrubs (15), Flowers (56), Grasses, Sedges and Rushes (19) and Ferns and fern allies (9). The respective figures for Ditchley are (8), (30), (10) and (1), and these include: Ramsons (*Allium ursinum*), Columbine (*Aquilegia vulgaris*), Wood Small-reed (*Calamagrostis epigejos*), Nettle-leaved Bellflower (*Campanula trachelium*), Midland Hawthorn (*Crataegus laevigata*), Spurge Laurel (*Daphne laureola*), Broad-leaved Helleborine (*Epipactis helleborine*), Wood Spurge (*Euphorbia amygdaloides*), Woodruff (*Galium odoratum*), Water Avens (*Geum rivale*), Toothwort (*Lathraea squamaria*), Narrow-leaved Everlasting-pea (*Lathyrus sylvestris*), Hairy Wood-rush (*Luzula pilosa*), Wood Millet (*Milium effusum*), Herb Paris (*Paris quadrifolia*), Greater Butterfly Orchid (*Platanthera chlorantha*), Polypody (*Polypodium vulgare*), Wild Service-tree (*Sorbus torminalis*), Betony (*Stachys officinalis*) and Wood Vetch (*Vicia sylvatica*).

Ditchley's total of 49 species out of a possible 99 is considered to be quite high. By comparison, Wychwood has 46, and 40 are common to both.

It would not have been possible to answer the third question, on the comparison between the floras of Ditchley and Wychwood, without the gift of a copy of the Wychwood species' list compiled several years ago by Peter Sheasby and his late wife Janet, members of the Botanical Society of the British Isles, who, over a period of nearly twenty years, in their spare time, recorded the flora of Wychwood for English Nature (Nature Conservancy Council). Their carefully compiled record was a most valuable and interesting source of reference during the Ditchley survey. As the recording seasons came and went, more and more of the plants on the Wychwood list were found on Ditchley's ancient woodland sites, until by the time the ultimate list of 443 species was prepared in 1992, 342 of Wychwood's 436 species (78%) were found to be common to both.

Among this special community of 342 plants are many which are now very scarce locally, including, to mention but a few, Lady's Mantle (*Alchemilla filicaulis* subsp. *vestita*), Columbine, Wild Liquorice (*Astragalus glycyphyllos*), Deadly Nightshade (*Atropa belladonna*), Harebell (*Campanula rotundifolia*), Spring Sedge (*Carex caryophyllea*), Meadow Saffron, Dropwort (*Filipendula vulgaris*), Common Rock-rose (*Helianthemum nummularium*), Horseshoe Vetch (*Hippocrepis comosa*), Toothwort, Common Gromwell (*Lithospermum officinale*), Common Restharrow (*Ononis repens*), Adder's-tongue (*Ophioglossum vulgatum*), Herb Paris, Common Milkwort (*Polygala vulgaris*) and Wild Service-tree. *Ranunculus trichophyllus,* the Thread-leaved Water-crowfoot, found in a small pool at Ditchley, is also the only member of the subgenus *Batrachium* to occur, in two pools, in Wychwood.

There are interesting differences too, some species occurring in Wychwood and not at Ditchley and *vice versa*. In some cases the presence or absence of a particular plant could be explained more or less satisfactorily, while in other instances the reasons were obscure. Wood Speedwell (*Veronica montana*), for example, is the commonest speedwell in Wychwood, but is rare at Ditchley, the likely explanation being that in general Wychwood is much damper. This probably also accounts for the fact that Wychwood has ten species of fern but Ditchley only five. Heather (*Calluna vulgaris*), Foxglove (*Digitalis purpurea*), Heath Bedstraw (*Galium saxatile*), Trailing St. John's-wort (*Hypericum humifusum*), Sheep's Sorrel (*Rumex acetosella*), Sessile Oak (*Quercus petraea*) and Heath Dog-violet (*Viola canina* subsp. *canina*) are plants of acidic, sandy soils, and they occur in Wychwood because there, on the highest ground, the Oxford Clay is capped by drift deposits of sand and gravel. Ditchley has no such deposits, so it is not surprising that these species were not recorded there.

On the other hand, it is difficult to explain why Spurge Laurel, Broad-leaved Helleborine, Woodruff, Wood Millet and Wood Vetch, all plants of oak-ash-hazel woods, have been found at Ditchley but not in Wychwood; but Wychwood alone has Green Hellebore (*Helleborus viridis*), Southern Marsh Orchid (*Dactylorhiza praetermissa*) and Violet Helleborine (*Epipactis purpurata*). And why Wychwood has only two or three but Ditchley more than sixty Wild Service-trees (the subject of a separate chapter) is even more difficult to understand.

At the end of this particular study, it was gratifying to have established a strong botanical link between Ditchley and Wychwood, and also to have shown that in spite of centuries of change, and their now irreversible separation, the two still share a flora that is not only remarkably similar but, compared to that of the neighbouring countryside, wonderfully rich.

RECORDS OF A CENTURY AGO

In the course of any botanical survey, in whatever part of the country it may be conducted, one of the most valuable, indeed indispensable, sources of reference is the county *Flora*. This is a book written by a botanist or botanists (usually with the prior assistance of a team of voluntary field workers) containing a record of all plants found growing wild in a particular county or region of Britain. Here, each species is given its English and Latin name, status (native or introduced), habitat, locality, distribution, frequency of occurrence and flowering period. Few parts of the British Isles are not covered by these highly interesting and informative publications.

In the case of Oxfordshire, until the new *Flora* is published in a few years' time, the handbook still in use is the important though now out-dated *Flora of Oxfordshire* by George Claridge Druce, which was published in 1886 and followed by a second, revised, edition in 1927. Oxfordshire being a large county, Druce divided it into seven districts based on the river systems. These were: (1) *Stour* (Severn), (2) *Ouse* (Ouse), (3) *Swere* (Upper Cherwell), (4) *Ray* (Lower Cherwell), (5) *Isis* (Upper Thames), (6) *Thame* (Mid-Thames), and (7) *Thames* (Lower Thames). On the map which accompanies the 1886 edition, Ditchley is seen as situated in *Isis*, the largest district, which extended from Heythrop in the north to Bampton in the south, and from Taynton in the west to Woodstock in the east; this is the region known today as West Oxfordshire.

It was interesting to work through the pages of both editions of the *Flora* listing the species Druce had recorded for Ditchley. There were eighty-nine. The question then was: how many had survived the intervening century, and if not all of them, then which ones? By the end of the survey, sometimes by chance and sometimes by direct search, sixty-six were rediscovered. These are marked '(GCD)' on the species' list (pp. 35–67). Because it is only in the county *Floras* that local plants are given specific locations, the list compiled from Druce's *Flora* is a good indicator of the most characteristic, uncommon or rare species recorded at that time.

The plants that have been lost evoke as much interest as those which have been found. It is easy, for instance, to explain the disappearance of Corncockle (*Agrostemma githago*), Corn Cleavers (*Galium tricornutum*) and Corn Buttercup (*Ranunculus arvensis*) from cornfields in Oxfordshire, because these are species which, over the past forty years or more, have become very rare or almost extinct nationally as a result of cleaner seed corn and changing agricultural practices. It is less easy to explain the disappearance of other species. Some may have been lost as a result of both drastic and subtle changes, wrought by man or the elements, which have destroyed or altered their habitats; delicate species could have succumbed to pressure from larger, more aggressive ones; a few with highly specialised requirements will have been unable to adjust to adverse changes in their environment; water plants may have fallen victim to predation by visiting Canada geese; and others would always have been vulnerable because their numbers were small or their methods of reproduction less successful. And there are plants which, once they have been lost from a particular place, never reappear.

Unfortunate though these losses are, it is a matter for deep satisfaction that so many of the species recorded by Druce are still flowering on the Ditchley lands. Among the most exciting rediscoveries were Columbine (*Aquilegia vulgaris*), Water Avens (*Geum rivale*), White Helleborine (*Cephalanthera damasonium*), Broad-leaved Helleborine (*Epipactis helleborine*) and Greater Butterfly Orchid (*Platanthera chlorantha*), the last three being the scarcest of the six different orchids which grow in a few of the rides and on edges of open woodland. The Water Avens, a delightful plant of marshes, damp woods, ditches and streamsides, is common in the north of England and Scotland but local or rare in the south. It has been recorded in only six of the total 596 tetrads (2 × 2 km squares) in Oxfordshire. The small colony at Ditchley is close to a spring which rises periodically in a hollow on the edge of a woodland ride. Some of the plants have crossed with the common yellow-flowered Wood Avens (*Geum Urbanum*). This happens frequently wherever the two species grow in close proximity, and the hybrids, which are easily recognised by their pale yellow instead of dusky orange-pink flowers, now outnumber pure specimens of Water Avens. The site is probably the same one reported to Druce by a Miss Pumphrey of Charlbury over sixty-five years ago.

An interesting plant which is not on the list because it was not refound is the wild Grape Hyacinth (*Muscari neglectum*), previously named *M. racemosum* or *M. atlanticum*. This species, first discovered by Sir John Cullum in Suffolk in 1776, is different from *M. armeniacum*, the Grape Hyacinth commonly grown in gardens. The flowers of the wild plant are almost entirely indigo blue, not bright blue, with white lobes. In the 1927 edition of the *Flora* Druce notes:

> [*Muscari racemosum* is] very abundant over a considerable portion of a large upland pasture near Ditchley Park where it has all the appearance of being native, and was in such quantity as to give colour to the field. This pasture during the War was tilled and much of it destroyed.

Even more intriguing was the recent discovery, in the Fielding-Druce Herbarium (Department of Plant Sciences, University of Oxford), of several pressed specimens of this *Muscari*, in Druce's collection, labelled simply:

Ditchley, Oxon. April 1931. J.Chapple with P.M.Hall.

On another sheet, in Druce's own hand is a note:

> *Muscari racemosum*. Upland pasture bordering Ditchley, Oxon. Native. May 1907.
> Capt. Gaskell [of Kiddington] & G.C. Druce.

Other specimens were from another upland field between Chadlington and Sarsden; and there was a further site at Chadlington, reported by a Miss

Burlton in 1911, where the plants were said to be '... in great abundance and native.' The *Muscari* still grows at Chadlington, and it was on a visit a few years ago that Lady Rosemary FitzGerald learnt from several of the older residents that, as children, they had earned a little pocket money by digging up and taking away the flowering bulbs, which were so plentiful as to be a 'nuisance' in the village allotments. But the exact locations of the sites at or near Ditchley, where once the *Muscari* grew, are likely to remain tantalizingly obscure.

Other local knowledgeable people who sent records, and possibly specimens, to Druce included Mr. J. M. Albright of the Charlbury Quaker family, Mr. H. Powell, and Lady Margaret Watney of Cornbury Park. How often the great Druce himself came to Ditchley to look for plants one cannot say, but from the convention, in the *Flora*, of placing an exclamation mark against the location of species he saw for himself, it is clear that he came to see Ramsons (*Allium ursinum*), White Helleborine, Meadow Saffron (*Colchicum autumnale*), Crested Hair-grass (*Koeleria macrantha*), Lesser Butterfly Orchid (*Platanthera bifolia*), and Greater Butterfly Orchid – all special plants which perhaps he had been told about. It must have been on one of these visits that he noticed the Common Valerian (*Valeriana officinalis*) growing 'plentifully' there, as it still does. And there is no doubt that he once picked a Lesser Butterfly Orchid, because a specimen in his herbarium at Oxford is labelled: 'Ditchley, Oxon. June 1884. G.C. Druce.' This is one of the species which appears to have been lost; but there is always the possibility that in corners of the estate as yet unexplored, this orchid and a few more of Druce's plants await rediscovery.

Because his *Flora* was a valuable source of reference during the survey, and because he was such a remarkable man, it seems fitting to end this chapter with a few notes on Druce's life. He was born in the village of Potterspury in Northamptonshire in 1850. After training as a pharmacist and working for a time in his home county, he settled in Oxford, conducting his business from a shop he owned in the High Street. A man of enormous energy and strong physique, and passionately interested in wild plants, he wrote three county *Floras* (Northamptonshire, Oxfordshire, and Berkshire) before the turn of the century, and while still a comparatively young man became known far beyond the bounds of Oxfordshire as an outstanding field botanist of wide experience. Active also in the business and political life of the city, he was elected Sheriff in 1897 and Mayor in 1900, spending forty years on the city council. Retiring at around the age of fifty, and by then a wealthy man, he devoted the greater part of his remaining years until he died at the age of eighty-two to travellling widely at home and abroad in pursuit of plants. He continued to collect specimens, adding these to his already extensive herbarium, at the same time acting as Special Curator of the Fielding Herbarium at Oxford and exercising a controlling hand in the affairs of the Botanical Exchange Club. Regarded as one of the most remarkable botanists of his day, he left a strong and long-lasting influence on the study of botany in this country.

THE WILD SERVICE-TREES

Of all the native trees that grow in Ditchley's woods and hedgerows the Wild Service (*Sorbus torminalis*) is the most interesting, and one of the loveliest. Although related to the Rowan and in the same genus as the Whitebeams, it is very different in appearance; and to this distinctiveness is added an air of mystery, for it was one of the trees of the original wildwood, or primary woodland, that formed after the last Ice Age and is therefore of truly ancient origin. It occurs in England as far north as Westmorland, southern Yorkshire and the Humber and is found throughout central, southern and western Europe, southwards as far as Algeria and eastwards to Denmark and the Caucasus. It is regarded as rare in England but is known to occur abundantly in certain areas of the south-east (notably Sussex and Kent), parts of the Midlands, and the Wye valley. Described by some writers as a shrub or small tree, under optimum conditions in open deciduous woodland, usually of Oak or Ash, it will grow to sixty or seventy feet; and at that height, with its heavy branches and spreading crown, it looks no less majestic than an oak. It is very scarce in Oxfordshire, having been recorded in only 30 of the county's 596 tetrads (2 × 2 km squares).

The silviculturalist and diarist John Evelyn (1620–1706) knew the Wild Service-tree well, and there is no better description of it than in his book *SILVA, or a Discourse of Forest-Trees and the Propagation of Timber* (1664), as the following passage shows:

> The Wild, or Maple-leaved Service . . . grows naturally in many parts of England, and is chiefly found upon strong soils . . . It rises to the height of forty or fifty feet, with a large trunk spreading at the top into many branches, so as to form a large head. The young branches are covered with a purplish bark, marked with white spots, and are garnished with leaves placed alternately, standing on pretty foot-stalks; these are cut into many acute angles, like those of the Maple-tree, and are near four inches long and three broad in the middle, having several smaller indentures toward the top, of a bright green on their upper sides but a little woolly on their under. The flowers are produced in large bunches toward the end of the branches; they are white, and shaped like those of the Pear-tree, but smaller, and stand upon longer foot-stalks; these appear in May, and are succeeded by roundish compressed fruit, which are shaped like large Haws, and ripen late in autumn, when they are brown.

In his 'Notes on the Wild Service Tree' (*International Dendrology Society Year Book 1982*), Mr. Patrick Roper, who for some years has been studying the distribution, ecology and economic history of this tree, writes:

> In Britain the Wild Service is scarce both in the wild and in cultivation. The most favoured habitat is mature natural forest, especially on clay soils and it does not occur above about 100 metres, or on marshy soil. . . . In what is left of our ancient forests today, I know of only three, widely separated, woods where probably more than 100 trees occur. Normally they grow singly or as a group of two or three, often far away from their nearest neighbour. . . . It would seem that the Wild Service was never very abundant in Britain although, no doubt, a good deal commoner than it is now.

Druce noted that the Wild Service was more plentiful in Wychwood before it was disafforested in 1853. Today, only two or three trees have been recorded in the remaining fragment of the Forest that is now a National Nature Reserve of about 1,500 acres.

What is so intriguing about the Wild Service-trees at Ditchley is their abundance. Although a few have been found on some of the old deciduous woodland sites and in hedges, their numbers are surprisingly large in Big Park. These 36 acres, enclosed for over 350 years in the deer park, were planted with beech and conifers in 1955 and 1958. At that time the whole area was covered with bracken and thorns and canopied mainly by large oaks. After an initial exploration in 1989, when twenty-one Wild Service-trees were recorded in one corner of the old parkland which is more open, in 1992 it was decided to make a thorough survey. This was done over several days in November and December, it being easier at that time of year to pick out the bare trees and, as the poet John Clare might have observed, their 'cranking' (crooked) boughs and characteristic grey-brown, rough, flaky bark. The total count was sixty-six, of which six were dead but still standing, and one hollow but alive and bearing fruit. Each tree was measured at the recommended height of five feet from ground level, then temporarily numbered, to make sure that none was overlooked or counted twice. The results showed that the trees are of mixed age, their circumferences ranging between 2 ft. 4 ins. (0.71 m) and 7 ft. 6 ins. (2.28 m). The total number of living trees found on the estate is sixty-seven. Such a large number is considered by Dr. Rackham to be exceptional. The head forester has reported about a dozen saplings growing well on one of the old woodland sites, and four were found in Big Park and a narrow belt of woodland elsewhere in the winter of 1992.

The measurements of the two largest trees were sent to Mr. Alan Mitchell, a well-known authority on trees and Director of the Tree Register of the British Isles. He calculated, from their girths, 7 ft. (2.13 m) and 7 ft. 6 ins. (2.28 m), that they are about 160 years old, these measurements, according to his records, being '. . . well beyond those of any [Wild Service] with a known date.' He also said: 'They are quite big while not in the outsize class . . . and also have very good stems' (clean trunks of up to 18 ft.), adding, '. . . the great feature is that this tree reproduces itself only in primary woodland. If the wood area has ever been ploughed, seedlings will not arise. It is a primary woodland indicator.'

It was noticed that several of the trees in the denser areas of the plantation are dying back, but although the lower branches are dead the crowns are still alive. The Wild Service is apparently tolerant of light shade in its early years but needs space and open sky above if it is to grow well, and if these requirements are absent then it becomes shaded out. In Big Park the finest trees are in more open situations or on the edges of wide rides. One truncated tree in a hedgerow has a spread of 36 ft. (11 metres).

Although the Wild Service produces abundant fruit, it does not grow successfully from seed, which may take two years to germinate. Many seeds are eaten by birds (including winter-visiting fieldfares), insects and grubs, or fall victim to decay. John Evelyn recommended that '. . . the best way is therefore to propagate them of suckers, of which they put forth enough.' They do indeed. Some of the Ditchley trees are encircled by a dense growth of these shoots, most of which grow close to the trunk, while a few have been located several feet away.

Mr. Ernest Lloyd, who studied the Wild Service in Epping Forest, has shown that regeneration is achieved principally by suckers which arise from long runner roots extending up to 110 metres from the parent tree. He also found that trees which have developed from these suckers have a curious habit of leaning over towards the parent tree. So it was interesting to find, in the south-western corner of Big Park, a group of 17 young trees of similar age, including a few with curved trunks, growing closer together than others do elsewhere; and between and among them at least a dozen well-rotted, moss-covered stumps of old Wild Service. Because the wet, decayed wood is a rich red-brown colour and tends to rot from the centre in a radial fashion, it was not difficult to distinguish these stumps from those of other trees. One was impressive, measuring almost four feet across, its moss-filled centre completely rotted away and sporting several green fronds of Broad Bucker-fern. It is possible that the young trees here have developed from suckers produced many years ago by these long dead trees and that this has happened elsewhere.

Because the fruits are attractive to pheasants it has been suggested that the Wild Service-trees in Big Park were planted to provide the birds with an additional source of wild food. Patrick Roper has learnt that the tree was '. . . definitely planted, or conserved in woodlands for pheasants from time to time, more especially in Germany than England.' In the case of Big Park, if many young trees had in fact been introduced simultaneously at some time in the past, they would almost certainly have grown to be more or less even-aged; but, as has been shown, there is great variation in the age of the trees. A different explanation may be that as this area was enclosed long ago within the deer park it was not subject to the same regime of coppicing and felling as were the woods outside it. Parkland has always been prized for its amenity and prestige value, the combination of elegant landscape and fine trees being one of its most attractive and enduring features. It could be that in Big Park the Wild Service- trees were left to regenerate naturally and have matured alongside other native species – the Field Maple, Crab, Oak and Beech – still to be found in this relatively undisturbed corner of the parkland formed centuries ago out of primary woodland. If this was so, then the fact that pheasants enjoy the fruits would have been incidental, but at the same time would have helped to preserve these wonderful trees.

If left to blet, as Medlars are, the brown fruits become soft and wrinkled, when they resemble sultanas. Although slightly gritty, they have a pleasant, mildly ascorbic flavour. In the past, the berries were used medicinally as a cure for colic, dysentry and other such conditions. (The word *'torminalis'* in *Sorbus torminalis* is from *'tormina'*, meaning 'colic'.) On their edibleness, Patrick Roper writes:

> In the Weald, particularly its Kentish section, berries were threaded by their stalks on to a stick with a snag at the end. This was then hung outdoors to get frosted, a process which hastens ripening . . . An elderly acquaintance tells me that each member of his family used to have a stick of Wild Services when he was a boy on a farm in Kent and that he used to eat just one or two berries a day throughout the winter. This self-imposed rationing, especially in a small boy, is interesting and it could be that the fruit, which is known to have a high vitamin C content, was traditionally and unconsciously used to help make up for the winter lack of fresh fruit and vegetables.

In the 17th century John Evelyn noted that the fruit '. . . was sold in the London markets in autumn . . . ', and in 1910 Druce recorded that the fruits from Wychwood trees were sold in Witney market, in west Oxfordshire.

Few trees can rival the Wild Service for autumn colour. The leaves turn yellow, orange, gold, red, scarlet, pale or dark brown, or remain partially green and are often flecked or blotched with a mixture of these hues. A few dozen of the most colourful ones collected from the woodland floor while still fresh, left to dry and then placed in a wooden bowl will delight the eye throughout the winter.

The Wild Service was one of Martin's favourite trees. He admired it in all seasons, but especially for its beauty in autumn. To ensure that this very special tree would always be found on the estate, in 1986 he ordered fifty young specimens for planting in newly created copses and plantations, and also in hedgerows, and three years later a further ten were set in the Pleasure Grounds. Fortunately for future researchers, details of these plantings are safely recorded in the head forester's day book.

WOODLAND GRASSLAND AND OLD PASTURE

WOODLAND GRASSLAND

In his masterly book *Ancient Woodland* Dr. Rackham draws attention to the importance of 'woodland grassland', a term he uses to describe the type of grassland which has always existed within the boundaries of a wood. Such grasslands are known to be prehistoric in origin, a fact which has been established by botanists who specialise in the study and identification of deposits of pollen and plant remains dating back to before the Atlantic Period (before 5500 B.C.). The most interesting thing about them is that they support plant communities which are distinctly different from those of the surrounding woodland.

Not long after the survey started a small but excellent example of this kind of grassland was found at Ditchley, in a clearing on an ancient woodland site now surrounded by plantations. Over sixty species were recorded there, including Lady's Mantle (*Alchemilla filicaulis* subsp. *vestita*), Daisy (*Bellis perennis*), Glaucous Sedge (*Carex flacca*), Common Centaury (*Centaurium erythraea*), Eyebright (*Euphrasia nemorosa*), Dropwort (*Filipendula vulgaris*), Lady's Bedstraw (*Galium verum*), Fairy Flax (*Linum catharticum*), Bird's-foot Trefoil (*Lotus corniculatus*), Field Wood-rush (*Luzula campestris*), Narrow-leaved Meadow-grass (*Poa angustifolia*), Common Milkwort (*Polygala vulgaris*), Tormentil (*Potentilla erecta*), Cowslip (*Primula veris*), Lesser Stitchwort (*Stellaria graminea*), Goat's-beard (*Tragapogon pratensis*), Hop Trefoil (*Trifolium campestre*), Yellow Oat-grass (*Trisetum flavescens*), Thyme-leaved Speedwell (*Veronica serpyllifolia*) and Hairy Violet (*Viola hirta*).

Also in this area were species which occur both in grassland and open woodland. Among these were Bugle (*Ajuga reptans*), Columbine (*Aquilegia vulgaris*), Common Spotted Orchid (*Dactylorhiza fuchsii*), Twayblade (*Listera ovata*), Adder's-tongue (*Ophioglossum vulgatum*), Early-purple Orchid (*Orchis mascula*), Greater Butterfly Orchid (*Platanthera chlorantha*) and Primrose (*Primula vulgaris*).

Additional species, growing in slightly disturbed ground, were the uncommon Deadly Nightshade (*Atropa belladonna*), Woolly Thistle (*Cirsium eriophorum*), Imperforate St. John's-wort (*Hypericum maculatum*), and Common Gromwell (*Lithospermum officinale*). Here, too, was Ploughman's Spikenard (*Inula conyzae*), found nowhere else on the estate.

A much smaller, triangular piece of grassland measuring only 15 × 18 × 18 ft. (4.56 × 5.49 × 5.49 m), and connected to the larger one by a narrow track, was found to contain no less than forty-two species. One of these, Devil's-bit Scabious (*Succisa pratensis*), has so far been found in only one other place. It is of special interest because, to quote Dr. Rackham further, it was one of the '. . . common plants in the tundra that preceded the wildwood [prehistoric forest of the Atlantic Period, 5500–3100 B.C.], for which their present association with rides or coppicing happens to provide a substitute.'

For thousands of years the short turf of these grasslands was preserved by the grazing herds of great beasts that roamed the primeval forests. By Anglo-Saxon times the free or controlled grazing of woodland (wood-pasture) by man's domesticated animals – mainly cattle and sheep – was a common practice which even today has not entirely disappeared. Also, of course, large areas of forest were enclosed, or imparked, to contain the royal deer. In many woods deer still roam freely today. They do at Ditchley, but there the single most important influence in preserving the 'grazed' conditions needed by the more delicate species, and preventing the grassland from reverting to scrub, is the estate's policy, as part of its woodland management, of 'swiping', or cutting, these areas, and all the rides, in late summer or early autumn each year.

In Dr. Rackham's opinion woodland grassland '. . . may be important not only as an element in the wood but also as a distinctive type of grassland, especially in areas where most ancient meadows and pasture have been destroyed or fertilized.' He also says: 'The disappearance of woodland grassland has been mentioned as the second most important loss to the flora of woods.'

The Ditchley examples are therefore of considerable interest and importance.

OLD PASTURE

The two unimproved pastures situated close to the south-eastern boundary of the estate are also of great botanical interest. They lie to the north and south of a narrow rectangular plantation which surrounds a rough, wet piece of ground known as 'The Bog'. Of the two, the larger field has the more varied terrain: its northern and central parts are slightly elevated and drier; its eastern border, which runs between 'The Bog' and a small pond, is permanently damp or wet; and there are two small dry banks colonised by a few unusual species which have been found nowhere else on the Ditchley lands.

Together, these pastures support a rich flora, of a type which has become increasingly rare. They are reminiscent of the old herb-rich calcareous grasslands which were a common sight in chalk and limestone country fifty or sixty years ago. Cowslips still flower here in profusion in early May – a rare and lovely sight.

Seventy-eight species were recorded, including sixteen different grasses and five sedges. Included in this list are Marsh Foxtail (*Alopecurus geniculatus*), Quaking-grass (*Briza media*), Harebell (*Campanula rotundifolia*), Cuckooflower (*Cardamine pratensis*), Spring Sedge (*Carex caryophyllea*), Distant Sedge (*C. distans*), Dwarf Thistle (*Cirsium acaule*), Common Spotted Orchid (*Dactylorhiza fuchsii*), Common Spike-rush (*Eleocharis palustris*), Meadow Fescue (*Festuca pratensis*), Common Rock-rose (*Helianthemum nummularium*), Downy Oat-grass (*Helictotrichon pubescens*), Oxeye Daisy (*Leucanthemum vulgare*), Fairy Flax (*Linum catharticum*), Common

Restharrow (*Ononis repens*), Adder's-tongue (*Ophioglossum vulgatum*), Mouse-ear Hawkweed (*Pilosella officinarum*), Burnet-saxifrage (*Pimpinella saxifraga*), Common Fleabane (*Pulicaria dysenterica*), Bulbous Buttercup (*Ranunculus bulbosus*), Salad Burnet (*Sanguisorba minor* subsp. *minor*), a microspecies of Dandelion (*Taraxacum oxoniense*) and Hairy Violet (*Viola hirta*).

Until about 1987 these two pastures, totalling approximately 14 acres, were regularly cut for hay. Since taken over by the estate they have been grazed between early April and late autumn by cattle, and occasionally by sheep. Mowing and/or grazing are essential if the rich flora of such grasslands is to remain balanced in number and variety of species. Many of the plants growing in them are not only palatable to livestock but contain valuable minerals and trace elements. In the past there were farmers who firmly believed that such herb-rich swards were healthy for cattle and sheep, and would reserve one such field, often a small one, for an ailing beast to graze in. A high percentage of these old permanent pastures were either ploughed during World War II to grow food or, since then, have been fertilized or ploughed and re-seeded with mixtures containing only a few grass species, rye-grass predominating. When this happens the unique plant communities of these grasslands vanishes for ever.

Remnants of woodland grassland and old pasture, wherever they exist in Britain, are botanical heirlooms representing a link with the original wildwood, and like all valuable heirlooms from a distant and richer past, they deserve our respect; and because we have lost so many of them, those that remain need to be preserved. It was because Martin was so keenly interested in and so admired the wild flowers of the estate that he gave instructions that the cowslip pastures were to remain unfertilized, and that care should be taken by the forestry staff to avoid damaging the woodland grassland when softwoods were extracted from the adjoining plantations. There can be no better example of simple but effective action for wild plant conservation.

Bole of the great beech near Ellen's Lodge.

10 May 1989

Looking north towards Little Park.

Winter 1989

Bluebells and young beech after removal of Douglas Fir and Norway Spruce between 1981/82 and 1988/89 from a 1954–56 plantation

30 April 1989

Primroses among poplars, with beech-larch plantation. The ground flora also includes Anemone(*Anemone nemorosa*), Yellow Archangel (*Lamiastrum galeobdolon*), Toothwort (*Lathraea squamaria*), Twayblade (*Listera ovata*), Dog's Mercury (*Mercurialis perennis*) and Common Dog-violet (*Viola riviniana*).

27 April 1991

Wood Vetch (*Vicia sylvatica*) on the edge of a 1964 plantation after extraction of Scots Pine in 1988.

21 June 1988

Early-purple Orchids (*Orchis mascula*) in a woodland clearing.

23 April 1988

Meadow Saffron (*Colchium autumnale*). Widespread in woodland rides and clearings.
19 August 1988

Nettle-leaved Bellflower (*Campanula trachelium*). One of the plants recorded for Ditchley by G.C. Druce a century ago.

21 July 1988

Autumn leaves and fruit of a young Wild Service-tree (*Sorbus torminalis*) in a hedge.
26 September 1990

Looking east from Sladhollow bank, Cleveley, towards Radford. In the foreground, old calcareous grassland supporting a variety of plants including Wild Liquorice (*Astragalus glycyphyllos*). Willow-lined River Glyme in middle distance, improved permanent pasture beyond.

26 March 1989

Cowslips in old pasture, where seventy-eight species were recorded. Plantation bordered with Wild Cherry (*Prunus avium*).

9 May 1989

Cornfield Flowers

The arable lands at Ditchley are extensive. They cover more than 3,000 acres and are farmed as three separate units in the manner common to all present-day efficient agricultural enterprises. The soil is described as 'good limestone brash, with clay pockets' and well suited to the production of cereals. The principal crops are wheat and barley, with a smaller acreage of oats, and 'break' crops of beans, peas, oilseed rape and, since 1991, flax (linseed).

The margins of standing crops and their autumn stubbles have always attracted botanists, who still find them interesting and sometimes rewarding hunting-grounds. Confronted, however, with so large an acreage, it was decided, at least for the time being, to explore only those fields immediately accessible from the estate roads, near farm buildings, or on the borders of wooded areas. This sample, although small, is probably representative of the cultivated land as a whole.

It so happened that during the period of the Ditchley survey the Botanical Society of the British Isles, concerned at the increasing scarcity, rarity or near-extinction of many once familiar cornfield flowers, selected 25 vulnerable species for survey by its members in 1986/87. After searching the selected fields over several years, eleven of these species have been found on the estate. They are: Corn Chamomile (*Anthemis arvensis*), Dwarf Spurge (*Euphorbia exigua*), Red Hemp-nettle (*Galeopsis angustifolia*), Sharp-leaved Fluellen (*Kickxia elatine*), Round-leaved Fluellen (*K. spuria*), Venus's-looking-glass (*Legousia hybrida*), Prickly Poppy (*Papaver argemone*), Corn Parsley (*Petroselinum segetum*), Night-flowering Catchfly (*Silene noctiflora*), Field Woundwort (*Stachys arvensis*) and Narrow-fruited Cornsalad (*Valerianella dentata*). There is great variation in their numbers. Only Dwarf Spurge and Round-leaved Fluellen are present in quantity. Night-flowering Catchfly, Sharp-leaved Fluellen and Venus's-looking-glass appear in a few fields each year, but their numbers are always small. Corn Parsley grows in varying amounts year by year in the same four widely separated places, all of which are at the edges of grass verges or tracks close to crops. Prickly Poppy (one plant) and Narrow-fruited Cornsalad (four plants) were rarities. The discovery, in 1989, of about 120 Field Woundwort around the perimeter of a bean stubble field was a surprise, as this is normally a species of non-calcareous soils; and since then a few more plants have been found in bean fields elsewhere on the estate.

Several Red Hemp-nettle were seen in 1983 along one side of an oilseed rape field on the north-west boundary. In July 1986 the count was 334 (the greatest number of Red Hemp-nettle recorded for the B.S.B.I. survey at any one site). Growing with them were 315 Dwarf Spurge, 6 Sharp-leaved Fluellen, 91 Round-leaved Fluellen, 31 Night-flowering Catchfly and 39 Narrow-fruited Cornsalad. By 1992 all these species had disappeared, with the exception of a few Dwarf Spurge.

Among the commoner cornfield flowers found on the cultivated lands are Parsley-piert (*Aphanes arvensis*), Small Toadflax (*Chaenorhinum minus*), Henbit Dead-nettle (*Lamium amplexicaule*), Corn Mint (*Mentha arvensis*), Field Madder (*Sherardia arvensis*), Green Field-speedwell (*Veronica agrestis*) and Field Pansy (*Viola arvensis*). Also Yellow-juiced Poppy (*Papaver dubium* susbsp. *lecoqii*), an uncommon poppy, small numbers of which occur sporadically in west Oxfordshire.

Two specimens of an alien plant, Common Amaranth (*Amaranthus retroflexus*), a native of N. America, were found in a fodder beet field in 1989. Seeds of this plant were probably among those of the beet at the time of sowing.

Many farmers, as well as botanists, will have observed that most of the wild plants growing on arable land are not perennials but annuals. These are of two types: summer annuals and winter annuals. The former develop from seeds which remain dormant throughout the winter and germinate in the spring or early summer, while the latter flower in the spring from seeds which have germinated the previous autumn. Yet others germinate, mature and disperse their seeds throughout the year if conditions are favourable.

In the course of their evolution many small, low-growing plants which we associate with cornfields have migrated from their original habitats into the heart of the wheat-fields, finding shelter there among the corn-stalks, at the same time avoiding competition from larger, more vigorous species. This strategy was highly successful in the days when farmers pulled weeds by hand, cut their corn earlier and stooked the sheaves to dry in the fields, and then left stubbles unploughed until the spring. This allowed time for a late summer or early autumn flush of annuals which quickly matured and dropped their seeds, adding these to the many already shed by other plants in previous months. A vast bank of seeds of many species was thus built up in the soil.

Efforts to clean the corn began with the simple process of hand-winnowing in a strong current of air; this separated at least a proportion of the lighter, smaller weed seeds from the grains of wheat, barley, oats and rye. Later, the threshing machine cleverly deposited the 'trash' on to a ground-sheet at one end and the corn into a bag at the other. Since then, more sophisticated methods of screening have almost completely eliminated unwanted seeds. And while this has pleased the farmer, it has impoverished the flora of the cornfields. Today, the herbicides which farmers everywhere are obliged to use kill not only undesirable plants such as Charlock, Wild Oat, Curled and Broad-leaved Dock, Black-grass, Redshank and Fat-hen, but also the harmless wild flowers growing among the crops. And the now widespread practice of ploughing stubbles in late summer and autumn destroys many of these annuals before they have time to set seed. However, they do have another strategy which helps to enhance their chances of survival. The seeds of the majority of them are very long-lived, those of the Common Poppy (*Papaver rhoeas*), for example, remaining viable for at least a century; while those of many others are known to remain dormant for decades. This being so, even small numbers of plants help to perpetuate the seed bank, which although less rich than in the past, remains a valuable depository awaiting disturbance by the plough.

Some cornfield flowers, however, have vanished altogether, or have become very rare or scarce. Among these are Pheasant's-eye (*Adonis annua*), Corncockle (*Agrostemma githago*), Cornflower (*Centaurea cyanus*) and Corn Buttercup (*Ranunculus arvensis*), together with the less conspicuous Thorow-wax (*Bupleurum rotundifolium*), Corn Cleavers (*Galium tricornutum*) and Spreading Hedge-parsley (*Torilis arvensis*). The beautiful Corncockle, for instance, whose generic name *Agrostemma* is a combination of two Greek words meaning 'field' and 'garland', must have adorned the cornfields of southern Europe for thousands of years, eventually reaching Britain, where it is not native, in shipments of corn from the Mediterranean and, in more recent times, from Russia. Farmers have always regarded it as a noxious weed, because if its seeds, which contain a poisonous glucoside, are present in milled corn they taint the bread. The herbalist John Gerard, writing of the Corncockle in *The Boke of husbandrie* in 1523, remarked: '. . . what hurt it doth among corne, the spoil unto bread, as well in colour, taste and unwholsomnes, is better known than desired.' Another detested plant was the tall, bushy Corn Buttercup, known by several country names, including 'Devil's Claws', 'Devil's Currycomb' and 'Hungerweed'. It was a serious pest in cornfields because its seeds, being large and spiny, tangled with the crops and stuck firmly to the fetlocks of working horses, the coats of dogs, and to clothing.

Almost without exception, the cornfield flowers of today are charming small annuals which offer very little or no threat to crops. So it is regrettable that in the process of eliminating undesirable wild plants from arable land they too are destroyed. (No-one who has marvelled at their evolution, or, with the aid of a hand-lens, been struck by their beauty, could ever dismiss them as 'weeds'!) Happily, at Ditchley arrangements have been made to leave a herbicide-free conservation headland along the side of the field where the Red Hemp-nettle grew, in the hope that it, and the plants associated with it, will return. If this can be achieved, and if numbers of all these threatened species can be increased and stabilised in even a few such places on farms countrywide, at a time when cereal production is being reduced and land set aside, this will help to rescue them from extinction not only locally but nationally.

BUTTERFLIES AND PLANTS

Martin was keenly interested in butterflies. He observed and recorded them on his walks around the estate and would have seen all the species listed below. Although several contributions were made by other people, the records for three of the most uncommon ones – the Painted Lady, Clouded Yellow and White Admiral – were his own. And since, as he would have known, butterflies require specific plants for their caterpillars to feed on, and favour certain flowers for their nectar, a list of those recorded at Ditchley, including the plants associated with them, is not inappropriate.

English name	*Latin name*	Caterpillar food plants
Brimstone	*Gonepteryx rhamni*	Buckthorn.
Comma	*Polygonia c-album*	Common Nettle, Hop, Elm, Red Currant.
Common Blue	*Polyommatus icarus*	Leguminous plants, including Bird's-foot Trefoil, Common Restharrow, Black Medick and Lesser Trefoil.
Gatekeeper	*Pyronia tithonus*	Grasses, especially Cock's-foot and Common Couch; also Bents, Fescues, and Smooth and Rough Meadow Grasses.
Green-veined White	*Pieris napi*	Cuckooflower, Hedge Garlic, Garlic Mustard, Watercress, Wild Mignonette.
Holly Blue	*Celastrina argiolus*	Flower buds of Holly, Ivy, Dogwood, Buckthorn, Spindle.
Large Skipper	*Ochlodes venata*	Cock's-foot and Sheep's Fescue.
Large White	*Pieris brassicae*	Leaves of Cabbage, Nasturtium, Wild Mignonette.
Marbled White	*Melanargia galathea*	Grasses, including Sheep's Fescue, Red Fescue, Cock's-foot, Cat's-tail, Tor-grass, Timothy.
Meadow Brown	*Maniola jurtina*	Various grasses, including Rye-grass and Smooth Meadow-grass.
Orange Tip	*Anthocharis cardamines*	Cuckooflower, Hedge Mustard.
Painted Lady	*Cynthia cardui*	Marsh, Musk and Spear Thistles, Mallow, Burdock.
Clouded Yellow	*Colias hyale*	Leguminous plants, particularly Clover and Lucerne.

English name	Latin name	Caterpillar food plants
Peacock	*Inachis io*	Common Nettle.
Purple Hairstreak	*Quercusia quercus*	Oak leaves.
Red Admiral	*Venessa atalanta*	Leaves and shoots of Common Nettle.
Ringlet	*Aphantopus hyperanthus*	Various grasses, including Cock's-foot, Common Couch and False Brome.
Small Copper	*Lycaena phlaeas*	Common Sorrel, Sheep's Sorrel and Dock.
Small Heath	*Coenonympha pamphilus*	Fine blades of grasses, such as Sheep's Fescue, Annual Meadow-grass and Bents.
Small Skipper	*Thymelicus sylvestris*	Soft-leaved grasses, including Yorkshire-fog.
Small Tortoiseshell	*Aglais urticae*	Terminal leaves of Common Nettle.
Small White	*Pieris rapae*	Brassicas, Nasturtium, Wild Mignonette.
Speckled Wood	*Pararge aegeria*	Grasses, mainly Cock's-foot and Common Couch; also False Brome and Yorkshire-fog.
Wall Brown	*Lasiommata megera*	Grasses, including Cock's-foot, Yorkshire-fog, Common Bent, False Brome, Tor-grass.
White Admiral	*Ladoga camilla* (*Limenitis camilla*)	Honeysuckle leaves.

The Purple Hairstreak and White Admiral are uncommon woodland butterflies, seldom seen even when present because of their tendency to keep to the tops of broad-leaved trees, particularly oak, where they often feed on honeydew deposited on leaves by aphids. The Painted Lady, Clouded Yellow and Red Admiral are regular spring and early summer migrants to Britain from their main breeding areas in the Mediterranean. Most of these butterflies are to be seen in open, sunny, flowery rides or clearings, or about hedgerows, while others seek dappled shade. Favourite nectar plants are Bramble, Clover, Field Scabious, Lesser and Greater Knapweed, Marjoram and Thistles. The Brimstone, Peacock and Small Tortoiseshell hibernate as butterflies, the Brimstone usually in thick ivy or holly, the other two more often in hollow trees.

The list was contributed to and compiled from records held by Mr. John Campbell, County Records Officer, Biological Records Centre, Woodstock.

REFERENCES

ALLEN, D.E. (1986) *The Botanists*. St. Paul's Bibliographies, Winchester.

CLAPHAM, A.R., TUTIN, T.G., & MOORE, D.M. (1987) *Flora of the British Isles*. Cambridge University Press.

DRUCE, G.C. (1886) *The Flora of Oxfordshire*. Parker & Co., Oxford.

DRUCE, G.C. (1927) *The Flora of Oxfordshire*. Second Edition. Clarendon Press, Oxford.

EMERY, F. (1974) *The Oxfordshire Landscape*. Hodder & Stoughton, Sevenoaks.

EVELYN, J. (1664) *SILVA, or a Discourse of Forest-Trees and the Propagation of Timber in his Majesty's Dominions*. York.

GRIGSON, G. (1975) *The Englishman's Flora*. Paladin, St. Albans.

JERMY, A.C., CHATER, A.O. & DAVID, R.W. (1982) *Sedges of the British Isles*. Handbook No. 1, Botanical Society of the British Isles, London.

KENT, D.H. (1992) *List of Vascular Plants of the British Isles*. Botanical Society of the British Isles, London.

LLOYD, E.G. (1977) *The Wild Service Tree Sorbus torminalis in Epping Forest*. The London Naturalist, No. 56.

MARREN, P.R. (1990) *Woodland Heritage*. David & Charles, Newton Abbot.

RACKHAM, O. (1980) *Ancient Woodland, its history, vegetation and uses in England*. Edward Arnold, London.

RACKHAM, O. (1986) *The History of the Countryside*. J.M. Dent & Sons, London.

ROPER, P. (1982) *Some notes on the Wild Service Tree, Sorbus torminalis*. International Dendrology Society Year Book 1982.

ROSE, F. (1981) *The Wild Flower Key*. Frederick Warne, London.

ROSE, F. (1989) *Grasses, Sedges, Rushes and Ferns of the British Isles and North-western Europe*. Viking, London.

SALISBURY, E. (1961) *Weeds and Aliens*. Collins New Naturalist, London.

SCHUMER, B. (1984) *The Evolution of Wychwood to 1400: Pioneers, Frontiers and Forests*. Leicester University Press.

STACE, C. (1991) *New Flora of the British Isles*. Cambridge University Press.

TANSLEY, A.G. (1949) *Britain's Green Mantle*. George Allen & Unwin, London.

PLANT SPECIES RECORDED BETWEEN 1985 AND 1992

(Wild flowers, grasses, sedges, rushes, ferns, shrubs and trees)

All Latin and English names follow the *New Flora of the British Isles*, by Professor Clive A. Stace (Cambridge University Press, 1991)

The list is arranged in Latin alphabetical order

Abbreviations:

(GCD) Recorded for Ditchley in G.C. Druce's *Flora of Oxfordshire* (1886, 1927)
(P) Planted

English Name	Latin Name	Notes
Field Maple (GCD)	*Acer campestre* L.	(Native) Mainly in hedgerows; several mature trees in open situations. The trunk of the largest tree has a girth of 10 ft. 6 ins. (3.20 m) at 3 ft. (0.91 m) from the ground, and divides into three tall stout branches at 3 ft. 4 ins. (1.10 m).
Norway Maple (P)	*A. platanoides* L.	(Introduced) Occasionally set at edges of plantations. Attractive yellowish-green flowers in early spring, before the leaves.
Sycamore (P)	*A. pseudoplatanus* L.	(Introduced from Central Europe; often self-seeds) In some plantations, but seldom conspicuous. Susceptible to squirrel damage.
Yarrow	*Achillea millefolium* L.	(Native; used medicinally from early times) On verges, banks and in clearings. Flowers late into the autumn.
Ground-elder	*Aegopodium podagraria* L.	(Believed to have been introduced by the Romans, or in the Middle Ages, as a pot-herb; *podagraria* means 'good for gout'.) Occasional patches near buildings.
Horse-chestnut (P)	*Aesculus hippocastanum* L.	(Introduced) Many fine specimen trees in the parkland and beside estate roads.

English Name	Latin Name	Notes
Fool's Parsley	*Aethusa cynapium* L.	(Native; a weed of cultivation) Infrequent on disturbed ground or edges of arable fields.
Agrimony	*Agrimonia eupatoria* L.	(Native; valued as a medicinal herb for many centuries) Infreqently scattered in clearings, on verges and by hedges.
Common Bent	*Agrostis capillaris* L. (*A. tenuis* Sibth.)	(Native; prefers acid grassland) Rarely recorded, but probably overlooked.
Black Bent	*A. gigantea* Roth	(Native; a grass of cultivated, rough or waste ground) Occurs rarely on arable land.
Creeping Bent	*A. stolonifera* L.	(Native) Here and there on damp verges or beside tracks.
Bugle	*Ajuga reptans* L.	(Native) Widespread in damp rides, shady places, and deciduous woodland areas.
Lady's-mantle	*Alchemilla filicaulis* subsp. *vestita* (Buser) Bradshaw	(Native) Occurs sparsely in old grassland on ancient woodland sites.
Water-plantain	*Alisma plantago-aquatica* L.	(Native) A few plants at either end of the lake.
Garlic Mustard	*Alliaria petiolata* (M.Bieb.) Cavara & Grande	(Believed to be native, but earliest record of Roman date) Widespread by hedges, but seldom plentiful.
Ramsons (GCD)	*Allium ursinum* L.	(Native) In several shady or damp woodland areas.
Alder	*Alnus glutinosa* (L.) Gaertner	(Native) Near the lake.
Marsh Foxtail	*Alopecurus geniculatus* L.	(Native; a perennial grass of wet places) Found in two places, beside streams.
Black-grass	*A. myosuroides* Hudson	(Native; a troublesome weed of arable land) Very little in cereal crops. Thousands of plants appeared after the planting of a new spinney in 1988, but subsequently disappeared. The seeds of this grass remain viable for only a few years.
Meadow Foxtail	*A. pratensis* L.	(Native; one of the first grasses to flower) Occasional in moist grassland.
Common Amaranth	*Amaranthus retroflexus* L.	(Introduced; native of N. America; transported in wool shoddy, soya-bean waste and bird-seed) Two plants in fodder beet crop, August 1989; probably arrived, as seed, with the fodder beet seed.
Scarlet Pimpernel	*Anagallis arvensis* L.	(Native; a charming, inoffensive 'weed' of arable land) Infrequent on cultivated land. Three blue-flowered plants found in oilseed rape stubble, July 1989.

English Name	Latin Name	Notes
Wood Anemone (GCD)	*Anemone nemorosa* L.	(Native) Scarce, and in small numbers in most deciduous woodland areas and at edges of plantations, often not flowering.
Wild Angelica	*Angelica sylvestris* L.	(Native; a robust plant of wet places) Widespread in damp grassland and dominant in two low-lying poplar plantations.
Barren Brome	*Anisantha sterilis* (L.) Nevski (*Bromus sterilis* L.)	(Native; a grass of waste places, sometimes troublesome on cultivated land) Found occasionally at edges of arable fields.
Corn Chamomile	*Anthemis arvensis* L.	(Native; an attractive, aromatic cornfield flower of chalk and limestone soils) Found in three places on arable land. A locally scarce species.
Sweet Vernal-grass	*Anthoxanthum odoratum* L.	(Native; contains the aromatic oil coumarin which gives the characteristic scent to new-mown hay) In permanent pasture, woodland grassland and on verges.
Cow Parsley	*Anthriscus sylvestris* (L.) Hoffm.	(Native; 'Queen Anne's Lace', or 'Keck') Common by hedgerows, along verges, etc.
Parsley-piert	*Aphanes arvensis* L.	(Native; an insignificant plant closely related to Lady's Mantle (*Alchemilla*). One of its country names was 'Breakstone', from its reputed ability to break stones in the kidney) Of rare occurrence on arable land.
Fool's Water-cress	*Apium nodiflorum* (L.) Lagasca	(Native) In a few places by pools and streams.
Columbine (GCD)	*Aquilegia vulgaris* L.	(Native) Rarely recorded in rides and woodland clearings. One plant appeared at the edge of a plantation after extraction of conifers, and several germinated and flowered in another plantation after several conifers were blown down in the winter gales.
Greater Burdock	*Arctium lappa* L.	(Native) Occasional on rough ground or edges of cultivated land.
Lesser Burdock	*A. minus* (Hill) Bernh.	(Native) Common and widespread in grassland, deciduous woodland areas, by tracks and on verges.
Thyme-leaved Sandwort	*Arenaria serpyllifolia* L.	(Native; a small plant of open ground, usually on light soils) Recorded in four places on or near cultivated land.
False Oat-grass	*Arrhenatherum elatius* (L.) P.Beauv. ex J.S. & C. Presl	(Native; a tall, common wayside grass) Widespread in rough grassland areas.
Mugwort	*Artemisia vulgaris* L.	(Native; a tall aromatic herb of hedgerows and verges) Rarely recorded in grassland.

English Name	Latin Name	Notes
Lords-and-Ladies or Cuckoo-pint	*Arum maculatum* L.	(Native) Frequent in old woodland areas and shady hedgerows.
Wild Liquorice (GCD)	*Astragalus glycyphyllos* L.	(Native; a local species of dry calcareous soils. Also known as Milk-vetch, in the belief that goats which ate the plant yielded more milk. The specific name *glycyphyllos* means 'sweet leaf') In several grassy places, in one of which it grows plentifully.
Common Orache	*Atriplex patula* L.	(Native; a weed of cultivated land) In varying quantity on arable land.
Deadly Nightshade	*Atropa belladonna* L.	(Native; a tall, poisonous plant of calcareous soils) Occurs occasionally in rides and clearings, or on disturbed ground. At the edge of one plantation, after conifer extraction in the winter of 1988/89, buried seed germinated. Of the 150 young plants recorded in the summer of 1989, only a few matured and flowered the following year.
Wild-oat	*Avena fatua* L.	(Introduced) Occurs rarely in arable crops.
Winter-cress	*Barbarea vulgaris* R.Br.	(Native) A few plants found in a wet ride in 1989.
Daisy	*Bellis perennis* L.	(Native) On verges, hedge-banks and in meadows.
Lesser Water-parsnip	*Berula erecta* (Hudson) Cov.	(Native) Surviving in one place only – a spring-fed pool which was filled in some years ago.
Silver Birch	*Betula pendula* Roth	(Native) Small numbers of mature and young trees on edges of rides, in woodland and plantations.
Tor-grass (GCD)	*Brachypodium pinnatum* (L.) P.Beauv.	(Native; a somewhat aggressive grass, an indicator plant of chalk and limestone grassland) In several places in clearings and rides, and on verges and banks.
False Brome	*B. sylvaticum* (Hudson) P. Beauv.	(Native; often a relic of old deciduous woodland) Occurs mainly along the borders of plantations on ancient woodland sites, or by shady hedgerows.
Quaking-grass	*Briza media* L.	(Native; now an uncommon species in west Oxfordshire) Frequent in unimproved permanent pasture and in one area of old calcareous grassland.
Upright Brome (GCD)	*Bromopsis erecta* (Hudson) Fourr. (*Bromus erectus* Hudson)	(Native; often dominant in old calcareous grassland) On grass verges and in unimproved pasture.

English Name	Latin Name	Notes
Hairy-brome	*B. ramosa* (Hudson) Holub (*Bromus ramosus* Hudson)	(Native; a tall, nodding grass of shady hedgerows and wood borders) Widespread and sometimes plentiful in most moist areas of old woodland.
Soft-brome	*Bromus hordeaceus* L. subsp. *hordeaceus* (*B. mollis* L.)	(Native; a common grass of waysides, grassy places and rough ground) Surprisingly scarce; recorded in only a few places.
Slender Soft-brome	*B. lepidus* O. Holmb.	(Origin uncertain; probably introduced) A few tufts of this small brome on an old grass track between arable fields.
White Bryony	*Bryonia dioica* Jacq.	(Native; a climbing plant with poisonous berries) Uncommon in hedges in various parts of the estate.
Box (P)	*Buxus sempervirens* L.	(Native only in certain localities such as Box Hill, Surrey, Boxley Hill on the North Downs, and Boxwell, Glos.) In a few places in the Pleasure Grounds. (It is interesting that on a Ditchley Estate map of 1726, in what is now Big Park, there were three areas marked 'Box Wood', 'Little Boxwood Piece' and 'Great Boxwood Piece'.)
Wood Small-reed	*Calamagrostis epigejos* (L.) Roth	(Native; a tall, attractive grass of damp woods and ditches, with purple-brown 'plumes') In a few places in damp rides adjoining old woodland areas.
Common Water-starwort	*Callitriche stagnalis* Scop.	(Native; ponds, ditches, or on wet mud) Recorded by one of the woodland pools.
Marsh-marigold	*Caltha palustris* L.	(Native) Groups of plants by the lake, ponds and streams.
Hairy Bindweed	*Calystegia pulchra* Brummitt & Heyw.	(Introduced; origin uncertain) Found climbing up nettles on waste ground; probably a garden escape.
Hedge Bindweed	*C. sepium* (L.) R.Br.	(Native) Rarely recorded.
Harebell	*Campanula rotundifolia* L.	(Native) In three widely separated places in old, short turf.
Nettle-leaved Bellflower (GCD)	*C. trachelium* L.	(Native; an indicator plant of ancient woodland) Significantly in varying numbers only in clearings and on borders of plantations on ancient woodland sites. Increasingly scarce in west Oxfordshire.
Shepherd's-purse	*Capsella bursa-pastoris* (L.) Medikus	(Native) Fairly common on cultivated land, by tracks and around farm buildings.

English Name	Latin Name	Notes
Hairy Bitter-cress	*Cardamine hirsuta* L.	(Native; a persistent weed of cultivation) Occasional on disturbed ground.
Cuckooflower	*C. pratensis* L.	(Native) Surprisingly scarce and in small quantity where it occurs by pools and streams, and in wet hollows.
Welted Thistle	*Carduus crispus* L. (*C. acanthoides* auct., non L.)	(Native; a tall, slender thistle with small purple flowers) Occasional in grassy places.
Musk or Nodding Thistle	*C. nutans* L.	(Native; prefers calcareous soils) In two places on or near arable land; one white-flowered plant.
Slender Tufted-sedge	*Carex acuta* L.	(Native; a sedge of 'ponds, dykes and riversides or of marshy places where there is a more or less constantly high water level') Plentiful in one of the woodland pools.
Lesser Pond-sedge	*C. acutiformis* Ehrh.	(Native) By the side of the lake.
Spring Sedge (GCD)	*C. caryophyllea* Latour.	(Native; a charming small early-flowering sedge of dry calcareous grassland) Recorded in only two places, in short turf of unimproved permanent pasture.
Distant Sedge	*C. distans* L.	(Native; usually by brackish and fresh-water marshes, mostly near the sea, but also occurs inland) One clump only, in unimproved permanent pasture, not far from marshy ground.
Grey Sedge	*C. divulsa* Stokes subsp. *divulsa*	(Native; a common sedge of hedgerows, verges and wood borders) In two or three places in the above habitats.
Leers' or Many-leaved Sedge	*C. divulsa* subsp. *leersii* (Kneucker) W. Koch	(Native; locally distributed and strongly associated with calcareous soils) In a few places in wet or damp rides.
Glaucous Sedge	*C. flacca* Schreber	(Native; the commonest sedge of calcareous grassland) Common and widespread in grassy places, particularly in old pasture.
Hairy Sedge	*C. hirta* L.	(Native) In damp grassland by streams or in the wetter parts of woodland rides.
False Fox-sedge	*C. otrubae* Podp.	(Native; wet places on heavy soils) In several places in this type of habitat.
Greater Pond-sedge	*C. riparia* Curtis	(Native) By one of the streams.
Bottle Sedge	*C. rostrata* Stokes	(Native) In quantity in one of the woodland pools.
Spiked Sedge	*C. spicata* Hudson	(Native) Recorded in one place, on a grass verge.

English Name	Latin Name	Notes
Wood-sedge	*C. sylvatica* Hudson	(Native) The commonest sedge on the estate. Widespread in almost all woodland areas and by shady tracks and rides.
Hornbeam	*Carpinus betulus* L.	(Native) One of the rarer trees at Ditchley. In a narrow belt of old deciduous woodland, possibly planted.
Sweet Chestnut (P)	*Castanea sativa* Miller	(Introduced) Another of the rarer trees at Ditchley; near the lake, and in a piece of old woodland.
Fern-grass	*Catapodium rigidum* (L.) C.E. Hubb. (*Desmazeria rigida* (L.) Tutin)	(Native; a stiff little grass of bare, dry soils) Many plants found in one place only, on a dry bank.
Common Knapweed	*Centaurea nigra* L.	(Native) Recorded in several places in rides and clearings. Uncommon.
Greater Knapweed	*C. scabiosa* L.	(Native) Recorded less often than the above species.
Common Centuary (GCD)	*Centaurium erythraea* Rafn	(Native; one of the most charming wild flowers of open, well-drained grassland) In a few locations in open, sunny rides where the grass is short.
White Helleborine (GCD)	*Cephalanthera damasonium* (Miller) Druce	(Native; an orchid of shady deciduous woods, usually beech, on chalk and limestone) Formerly plentiful on an ancient woodland site, now a plantation, where beech predominated. Only seven flowering plants were found in 1989 and four in 1990. It is hoped that numbers will increase after conifer extraction. A rare species in west Oxfordshire.
Common Mouse-ear	*Cerastium fontanum* Baumg.	(Native; a common species) In most grassy places.
Sticky Mouse-ear	*C. glomeratum* Thuill.	(Native) Rarely recorded, in grassland.
Small Toadflax (GCD)	*Chaenorhinum minus* (L.) Lange	(Native; a small annual of arable and waste land) At the edges of several arable fields.
Rough Chervil	*Chaerophyllum temulum* L. (*C. temulentum* L.)	(Native; an umbellifer related to but less common than Cow Parsley, and flowering later) In a few places by hedges, tracks and edges of plantations.
Lawson's Cypress (P)	*Chamaecyparis lawsoniana* (A. Murray) Parl.	(Introduced from the western United States) Widely used as a nurse tree for beech in the plantations.

English Name	Latin Name	Notes
Rosebay Willowherb (GCD)	*Chamerion angustifolium* (L.) Holub (*Epilobium angustifolium* L.)	(Native) Fairly common, but not in great quantity, in some woodland areas.
Fat-hen	*Chenopodium album* L.	(Native; a common weed of cultivated and waste ground) Occasional on arable land and around farm buildings.
Many-seeded Goosefoot	*C. polyspermum* L.	(Native) Several plants on the edge of a wheatfield in 1989 and in linseed (flax) crop in 1991.
Red Goosefoot	*C. rubrum* L.	(Native; often associated with manure heaps on farms) In a few places on waste soil near farm buildings.
Enchanter's-nightshade	*Circaea lutetiana* L.	(Native) Occurs sparingly in shady places in open woodland.
Dwarf Thistle	*Cirsium acaule* (L.) Scop.	(Native; a normally stemless thistle with spiny leaves; because often hidden in the grass, it is also known as 'the picnicker's thistle') In old grassland in five separate locations.
Creeping Thistle	*C. arvense* (L.) Scop.	(Native) In varying quantity on field edges and tracks, in clearings and occasionally in woodland. An unwelcome perennial weed whose flowers are attractive to butterflies.
Woolly Thistle (GCD)	*C. eriophorum* (L.) Scop.	(Native; a tall, handsome thistle of calcareous soils, local in distribution) Appears sporadically in clearings and rides, and abundantly on an old grassland bank.
Marsh Thistle	*C. palustre* (L.) Scop.	(Native; a tall, attractive thistle of marshes, damp woods and wet grassland) In numerous typical places on the estate. One specimen found in a wood was 9 ft. (2.74 m) tall.
Spear Thistle	*C. vulgare* (Savi) Ten.	(Native) In most grassland areas.
Traveller's-joy or Old Man's Beard	*Clematis vitalba* L.	(Native) Mainly in hedgerows and along woodland borders. Less frequent than expected, but can be troublesome in young plantations.
Wild Basil	*Clinopodium vulgare* (L.)	(Native; a slightly aromatic herb of woodland edges, grassland, scrub, hedgerows) In several of the rides and clearings.
Meadow Saffron (GCD)	*Colchicum autumnale* L.	(Native; sometimes called 'Autumn Crocus', but a member of the lily family. A local species of central and southern Britain flowering in late summer in open woods and damp meadows) One of the most beautiful wild flowers of the estate. In all the rides and edges of plantations

English Name	Latin Name	Notes
		on ancient woodland sites, sometimes in abundance, particularly where bracken and other dominant species such as Meadowsweet and Wild Angelica have been cut in the rides before the flowers appear in late August/early September. Recorded over a hundred years ago by Druce as 'very abundant in Wychwood and Ditchley Wood.'
Hemlock	*Conium maculatum* L.	(Native; poisonous) Recorded in only two places.
Pignut	*Conopodium majus* (Gouan) Loret	(Native; a delicate umbellifer of mildly acid soils) Occurs sparsely in some open woodland and grassland areas.
Field Bindweed	*Convolvulus arvensis* L.	(Native) Edges of tracks and cultivated land, but nowhere abundant. Regarded as a noxious weed, but redeemed by its lovely pink and white flowers.
Canadian Fleabane	*Conyza canadensis* (L.) Cronq. (*Erigeron canadensis* L.)	(Introduced) Not recorded until 1992, along edges of two arable fields.
Dogwood	*Cornus sanguinea* L.	(Native) Widespread in hedges, and on borders of woodland and plantations.
Swine-cress	*Coronopus squamatus* (Forsskaol) Asch.	(Native; usually on well trampled ground such as farm gateways and tracks) Several plants near old manure heap and on waste soil from fodder beet crop.
Hazel	*Corylus avellana* L.	(Native) Common and widespread in hedges, remnant deciduous woodland and on plantation edges. (The earlier woodlands were predominantly oak standards over hazel coppice.)
Midland Hawthorn	*Crataegus laevigata* (Poiret) DC.	(Native; a woodland rather than a hedge or scrub hawthorn, more tolerant of shade, and much less common. An ancient woodland species) In a few hedges and borders of old woodland.
Hawthorn	*C. monogyna* Jacq.	(Native) Common and widespread in hedges and occasionally along woodland edges.
Smooth Hawk's-beard	*Crepis capillaris* (L.) Wallr.	(Native) Occasional in grassland, including verges and banks.
Beaked Hawk's-beard (GCD)	*C. vesicaria* L. subsp *taraxacifolia* (Thuill). Thell. ex Schinz & Keller	(Introduced from Europe, first recorded in 1715 and now common in central and southern England) In several places in grassland and on disturbed and waste ground.
Crosswort (GCD)	*Cruciata laevipes* Opiz (*Galium cruciata* (L.) Scop.)	(Native; an uncommon plant in west Oxfordshire) Recorded in three places in damp grassland.

English Name	Latin Name	Notes
Crested Dog's-tail	*Cynosurus cristatus* L.	(Native. Druce noted (1927) that the spikes of this attractive grass were 'twisted into formal bouquets by village children') Fairly common in grassland.
Cock's-foot	*Dactylis glomerata* L.	(Native) Common in grassland and dominant in one ride in 1988.
Common Spotted-orchid	*Dactylorhiza fuchsii* (Druce) Soó	(Native) A fine display in one of the rides in 1987 (about 300 plants). In several damp rides, borders of old woodland and by the lake.
Spurge-laurel (GCD)	*Daphne laureola* L.	(Native; an evergreen woodland shrub whose green flowers open in February and are followed by black poisonous berries) In several open woodland areas.
Thorn-apple	*Datura stramonium* L.	(Introduced; an American plant which sometimes occurs on waste and cultivated ground, particularly in hot summers; poisonous in all its parts) Four plants appeared in one of the rides after soil was disturbed during extraction of Corsican Pines and Douglas Firs from an adjoining plantation.
Tufted Hair-grass	*Deschampsia cespitosa* (L.) P. Beauv.	(Native; a tall, elegant grass of damp meadows and woods, usually on heavier soils) Common and widespread in all damp grassland and woodland areas, where it is sometimes dominant.
Wild Teasel	*Dipsacus fullonum* L.	(Native) Occasional by tracks, verges and clearings. One of the favourite seed plants of the goldfinch.
Leopard's-bane	*Doronicum pardalianches* L.	(Introduced; native of western Europe, formerly cultivated for medicinal purposes) Extensive patches in one area of amenity woodland where it may have been introduced as a ground cover plant.
Broad Buckler-fern	*Dryopteris dilatata* (Hoffm.) A. Gray	(Native) Recorded in a few open woodland areas; less frequent than the following species.
Male-fern	*D. filix-mas* (L.) Schott	(Native) The commonest fern at Ditchley, and plentiful in a few clearings.
Common Spike-rush	*Eleocharis palustris* (L.) Roemer & Schultes	(Native) In two places near ponds.
Canadian Water-weed (GCD)	*Elodea canadensis* Michaux	(Introduced into Britain in 1836. It spread rapidly and is now naturalised in 'slow-flowing fresh waters' throughout most of Britain) In the lake.

English Name	Latin Name	Notes
Bearded Couch	*Elymus caninus* (L.) L.	(Native; a refined and harmless relative of the much-disliked Common Couch of cultivated land; a grass of woodland edges and hedgerows) Occasional by wood borders and shady hedgerows; probably under-recorded.
Common Couch	*Elytrigia repens* (L.) Desv. ex Nevski (*Elymus repens* (L.) Gould)	(Native) Widespread, and usually at edges of cultivated land, where it is not a serious problem.
American Willowherb	*Epilobium ciliatum* Raf. (*E. adenocaulon* Hausskn.)	(Introduced; a native of N. America first recorded in Britain in 1891 and now widespread) In a few places on disturbed ground.
Great Willowherb	*E. hirsutum* L.	(Native) Recorded in most damp woodland areas, by ponds, on wet banks and in damp hollows.
Broad-leaved Willowherb	*E. montanum* L.	(Native; in woods, hedgerows, waste places and often a weed in gardens) Seldom recorded in disturbed or bare ground.
Hoary Willowherb (GCD)	*E. parviflorum* Schreber	(Native; a willowherb of wet places) One of the scarcer plants, growing by pools and a stream.
Broad-leaved Helleborine (GCD)	*Epipactis helleborine* (L.) Crantz	(Native; commonest of the woodland helleborine orchids. Rare at Ditchley and now scarce in west Oxfordshire) Four flowering plants found in one of the plantations on an ancient woodland site in 1988. Long-established before the replanting in 1961, and it is hoped that numbers will increase after the conifers are removed.
Field Horsetail	*Equisetum arvense* L.	(Native) In a few damp places by pools and streams.
Marsh Horsetail	*E. palustre* L.	(Native) Rare; a few stems at the edge of a stream-fed pond.
Great Horsetail	*E. telmateia* Ehrh.	(Native; the largest and most striking of the British horsetails) Plentiful in rough vegetation beside one of the streams.
Common Whitlowgrass	*Erophila verna* (L.) DC.	(Native; not a grass, but a small ephemeral, winter annual of the Crucifer (Cress) family commonly found in dry, bare soil) In quantity under a wall near farm buildings.
Spindle (GCD)	*Euonymus europaeus* L.	(Native; a shrub or small tree valued for its attractive deep pink fruits which open out to reveal bright orange seeds) In moderate numbers in old hedges and woodland borders.

English Name	Latin Name	Notes
Hemp-agrimony	*Eupatorium cannabinum* L.	(Native) Recorded in only one place, at the side of a stream.
Wood Spurge (GCD)	*Euphorbia amygdaloides* L.	(Native; one of the most attractive spurges, flowering with the bluebells; an ancient woodland species) Occurs sparingly, or sometimes abundantly, in all the open areas of plantations on ancient woodland sites.
Dwarf Spurge	*E. exigua* L.	(Native; a small, now scarce, arable species) Widespread on cultivated land, where it is not a threat to crops.
Sun Spurge	*E. helioscopia* L.	(Native; a common spurge of cultivated ground) Rarely recorded.
Petty Spurge	*E. peplus* L.	(Native; a common weed of cultivation) Rarely recorded.
Eyebright	*Euphrasia nemorosa* (Pers.) Wallr	(Native; described as 'a highly critical genus', with about 20 species and 60 hybrids. A small, pretty plant still used as a remedy for tired or inflamed eyes) Recorded in two woodland grassland areas.
Beech (P)	*Fagus sylvatica* L.	(Native) The most important tree in the plantations. Numerous fine old trees in the parkland, by estate roads and hedges.
Black-bindweed	*Fallopia convolvulus* (L.) Á. Löve (*Polygonum convolvulus* L.)	(Native; a weed of arable land, waste places and gardens) In a few places on cultivated or waste ground.
Tall Fescue	*Festuca arundinacea* Schreber	(Native; a tall, stout grass previously used in some parts of Britain as a pasture grass) In one place on a wood border.
Giant Fescue	*F. gigantea* (L.) Villars	(Native; a very tall, shiny-leaved grass of woodland edges and shady places) At edges of clearings in old woodland areas, and dominant in one of the poplar plantations in 1988.
Sheep's-fescue	*F. ovina* L. subsp. *ophioliticola* (Kerguélen) M. Wilk.	(Native; described as 'a locally common subspecies of "grassy places on well-drained calcareous soils"') A few tufts on a grass track between arable fields.
Meadow Fescue	*F. pratensis* Hudson	(Native) Uncommon, in old grassland.
Red Fescue	*F. rubra* L. subsp. *rubra*	(Native) Widespread in grassy places.
Meadowsweet	*Filipendula ulmaria* (L.) Maxim.	(Native) Abundant and widespread, and often dominant in damp rides and near streams.

English Name	Latin Name	Notes
Dropwort	*F. vulgaris* Moench	(Native; a most attractive plant now very scarce in west Oxfordshire) In one small area of woodland grassland and in a remnant of permanent pasture.
Wild Strawberry	*Fragaria vesca* L.	(Native) Recorded in all the open woodland areas or adjoining clearings.
Ash	*Fraxinus excelsior* L.	(Native) Planted in the past, but not since 1953 as it does not do well on Ditchley land. Numerous mature and young trees and self-sown saplings. A few pollarded trees in the bottom of a meadow.
Common Fumitory	*Fumaria officinalis* L. subsp. *officinalis*	(Native) Appeared at Ellen's Lodge in 1991 after a chicken-run was dismantled.
Red Hemp-nettle (GCD)	*Galeopsis angustifolia* Ehrh. ex. Hoffm.	(Native; an attractive, uncommon annual of arable land) 334 plants recorded along one side of a wheat field in 1986, but none in 1992.
Common Hemp-nettle	*G. tetrahit* L.	(Native; often on arable land) Occasionally found at the edges of bean and oilseed rape fields.
Shaggy-soldier	*Galinsoga quadriradiata* Ruiz Lopez & Pavón (*G. ciliata* (Raf.) S.F. Blake)	(Introduced; a native of S. America; naturalised on waste and cultivated ground) One small plant in the middle of a track; unlikely to persist.
Cleavers	*Galium aparine* L.	(Native) Common and widespread, mainly in hedgerows.
Hedge Bedstraw	*G. mollugo* L. subsp. *mollugo*	(Native) Uncommon by hedges and at edges of clearings.
Woodruff (GCD)	*G. odoratum* (L.) Scop.	(Native; local, in woods on damp calcareous soils; an indicator plant of ancient woodland) This delightful plant occurs on the open borders of most plantations on ancient woodland sites, sometimes in abundance. Absent from Wychwood.
Common Marsh-bedstraw	*G. palustre* L.	(Native) By the lake and one of the ponds.
Lady's Bedstraw	*G. verum* L.	(Native; an attractive yellow-flowered bedstraw) In several widely separated places, always in small numbers.
Cut-leaved Crane's-bill	*Geranium dissectum* L.	(Native) Fairly common in grassland and along edges of cultivated fields.
Dove's-foot Crane's-bill	*G. molle* L.	(Native) In similar habitats, but less often recorded than the above species.

English Name	Latin Name	Notes
Meadow Crane's-bill (GCD)	*G. pratense* L.	(Native; a beautiful wild flower, characteristic of roadside verges in west Oxfordshire) On several tracks and verges, flowering profusely in a few of these places.
Hedgerow Crane's-bill	*G. pyrenaicum* Burman f.	(Possibly native; another charming crane's-bill) Found in only one place, in an old cattle yard.
Herb-Robert	*G. robertianum* L.	(Native) Fairly common in hedgebanks and open woodland areas.
Water Avens (GCD)	*Geum rivale* L.	(Native; a perennial plant of marshes, streamsides, ditches, damp open woods and mountain rock-ledges. More common in northern Britain, very local in the south) A small colony in a woodland ride, near a seasonal spring. One of Ditchley's special plants, not refound until 1990.
Water Avens × Wood Avens (GCD)	*G. rivale* × *G. urbanum* (*G.* × *intermedium* Ehrh.)	(Native; this hybrid is not uncommon, especially in the north of Britain, wherever the two species grow in close proximity) In 1991 about 20 hybrids were recorded, roughly twice the number of pure *G. rivale*. In the 1927 edition of the *Flora of Oxfordshire* Druce refers to this hybrid as 'rare' and records that a Miss Pumphrey found it 'In a wood near Ditchley Lodge'. This is probably the same site. It is pleasing to have discovered that after sixty-five years these pure and hybrid plants are holding on.
Wood Avens	*G. urbanum* L.	(Native) Common in all open woodland areas, edges of clearings, by tracks and hedges. (See above for hybrid between Water Avens and Wood Avens)
Ground-ivy	*Glechoma hederacea* L.	(Native) Common in damp woodland, grassland and hedgebanks.
Floating Sweet-grass	*Glyceria fluitans* (L.) R.Br.	(Native; a lush grass of marshes, rivers, ponds and wet meadows; relished by cattle) In a wet hollow in a woodland ride.
Plicate Sweet-grass	*G. notata* Chevall. (*G. plicata* (Fries) Fries)	(Native; in similar places to the above species) By two pools and a stream.
Common Ivy	*Hedera helix* L.	(Native) Widespread, but nowhere a menace. Rarely on the woodland floor.
Common Rock-rose (GCD)	*Helianthemum nummularium* (L.) Miller	(Native; a lovely wild flower, scarce in west Oxfordshire) On several grass verges, beside tracks and fields, and in old permanent pasture.
Meadow Oat-grass	*Helictotrichon pratense* (L.) Besser	(Native; an attractive grass of chalk and limestone soils) In remnant permanent pasture near a stream.

English Name	Latin Name	Notes
Downy Oat-grass (GCD)	*H. pubescens* (Hudson) Pilger	(Native) Recorded in three areas of old unimproved pasture.
Hogweed	*Heracleum sphondylium* L.	(Native; one of the commonest hedgerow and roadside plants) Recorded in all parts of the estate.
Dame's-violet	*Hesperis matronalis* L.	(Introduced into Britain as a garden plant, often escaping into the wild) In a few grassy places.
Horseshoe Vetch (GCD)	*Hippocrepis comosa* L.	(Native; an uncommon vetch of dry, calcareous grassland) A small colony discovered in 1992 on a low, dry bank in a pasture grazed by cattle.
Mare's-tail (GCD)	*Hippuris vulgaris* L.	(Native) Plentiful in two pools and in small quantity in the lake.
Yorkshire-fog	*Holcus lanatus* L.	(Native; one of the commonest grasses) In almost all damp grassland areas.
Creeping soft-grass	*H. mollis* L.	(Native; mostly on acid soils in woods, hedgerows) Recorded in one place in a woodland ride.
Bluebell (GCD)	*Hyacinthoides non-scripta* (L.) Chouard ex Rothm.	(Native) One of the glories of the estate, flowering abundantly and extensively in some places on the ancient woodland sites, especially after extraction of conifers from the plantations; also in old woodland areas previously coppiced and not replanted. Occasional white-flowered plants occur.
Hairy St. John's-wort	*Hypericum hirsutum* L.	(Native; mainly on basic soils) Frequent in damp grassland, woodland borders and rides.
Imperforate St. John's-wort	*H. maculatum* Crantz subsp. *obtusiusculum* (Tourlet) Hayek	(Native. More local than either the Hairy or Perforate St. John's-wort; prefers damp, heavy soils) In several damp rides and clearings.
Perforate St. John's-wort	*H. perforatum* L.	(Native) Frequent and widespread, often growing with Hairy St. John's-wort.
Square-stalked St. John's-wort	*H. tetrapterum* Fries	(Native; a *Hypericum* of watersides and wet grassy places) By lake, pond and streams.
Cat's-ear	*Hypochaeris radicata* L.	(Native; a common grassland plant with small dandelion-like flower heads) In clearings, old pasture and on grassy tracks.
Holly (GCD)	*Ilex aquifolium* L.	(Native; justly described by the diarist John Evelyn as an 'incomparable tree') Scattered in hedges, open woodland, rides and corners of fields. In a chapter on the geology of west Oxfordshire in a Geological Survey Memoir by H.B. Woodward, published in 1894, mention is made of the '... fine holly trees about Ditchley.'

English Name	Latin Name	Notes
Ploughman's-spikenard (GCD)	*Inula conyzae* (Griess.) Meikle	(Native; a local plant of open scrub and wood borders on calcareous soils) One fine group of plants in a woodland clearing.
Yellow Iris (P)	*Iris pseudacorus* L.	(Native) A few planted beside a stream.
Walnut (P)	*Juglans regia* L.	(Introduced) In a strip of woodland bordering an estate road.
Jointed Rush	*Juncus articulatus* L.	(Native) By ponds, stream and lake or in wet hollows.
Toad Rush	*J. bufonius* L.	(Native; a small annual rush often found on wet muddy tracks) By ponds and in a few rides where rushes grow.
Compact Rush	*J. conglomeratus* L.	(Native) In a wet ride and by a stream. Less common than the other rushes.
Soft Rush	*J. effusus* L.	(Native) In the wetter clearings and damp hollows.
Hard Rush	*J. inflexus* L.	(Native) Often found with Soft Rush, in the same type of habitat.
Sharp-leaved Fluellen	*Kickxia elatine* (L.) Dumort.	(Native; rather local, usually on calcareous soils) One of the 25 scarce arable plants selected by the Botanical Society of the British Isles for survey in 1986/87. Recorded in varying quantity in wheat, bean and oilseed rape fields, but less common than Round-leaved Fluellen.
Round-leaved Fluellen (GCD)	*Kickxia spuria* (L.) Dumort.	(Native; a delightful, rather local plant of calcareous soils) Also selected by the B.S.B.I. for their 1986/87 survey (see above). Almost always found with Sharp-leaved Fluellen, but in greater numbers. Plentiful in a bean stubble field in 1992.
Field Scabious	*Knautia arvensis* (L.) Coulter	(Native; a familiar plant of calcareous grassland) Scattered in various grassy places, always sparingly.
Crested Hair-grass (GCD)	*Koeleria macrantha* (Ledeb.) Schultes (*K. cristata* auct., non (L.) Pers.)	(Native; one of the less common grasses) Several tufts in short turf on a dry bank.
Prickly Lettuce	*Lactuca serriola* L.	(Probably native; usually found by waysides and on waste ground) Several plants on a heap of earth left after cleaning fodder beet.
Yellow Archangel (GCD)	*Lamiastrum galeobdolon* (L.) Ehrend. & Polatschek	(Native; associated with ancient woodland) In all the open areas and borders of plantations on the ancient woodland sites.

English Name	Latin Name	Notes
White Dead-nettle	*Lamium album* (L.)	(Native; common by hedges, on verges, edges of fields, flowering throughout the year) Recorded in almost all corners of the estate.
Henbit Dead-nettle	*L. amplexicaule* L.	(Native; uncommon, usually on cultivated ground on lighter, drier soils) In several cultivated fields, always in small numbers.
Red Dead-nettle	*L. purpureum* L.	(Native; very common on cultivated land and in waste places) Infrequent, mainly at edges of arable fields and farm tracks.
Nipplewort	*Lapsana communis* L.	(Native; waysides, wood margins, hedgerows) Common, sometimes as a weed in crops.
European Larch (P)	*Larix decidua* Miller	(Introduced) In many of the plantations, attractive at all times of year.
Toothwort	*Lathraea squamaria* L.	(Native; an unusual plant. One of its country names was 'corpse-flower', from its pale appearance and habit of arising from the naked earth in early spring. Associated with ancient woodland, and parasitic on woody plants, especially hazel) Thirty flowering spikes recorded in a piece of old hazel coppice in 1987, and a few in old woodland areas elsewhere.
Meadow Vetchling	*Lathyrus pratensis* L.	(Native) Widespread, but not plentiful, in grassy places and by hedges.
Narrow-leaved Everlasting-pea	*L. sylvestris* L.	(Native; a very attractive plant of scrub and wood borders; scarce in west Oxfordshire) Recorded in one small area of a clearing on the edge of a plantation on an ancient woodland site.
Venus's-looking-glass (GCD)	*Legousia hybrida* (L.) Delarbre	(Native; one of the most charming of all the small cornfield flowers, its name being derived from its shiny oval seeds '. . . like brilliantly polished brass mirrors'.) Occurs sparingly in some of the arable fields each year.
Common Duckweed	*Lemna minor* L.	(Native) In a corner of the lake, and in water-filled holes made by hooves of cattle in an old shallow pool.
Autumn Hawkbit	*Leontodon autumnalis* L.	(Native) Occasional in grassland.
Rough Hawkbit	*L. hispidus* L.	(Native; a dandelion-like plant with smaller, deep yellow flowers; usually in calcareous grassland) Occasional in grassy places.

English Name	Latin Name	Notes
Lesser Hawkbit	*L. saxatilis* Lam. (*L. taraxacoides* (Villars) Mérat)	(Native; less common than the above two hawkbits) A few in one area of unimproved permanent pasture.
Oxeye Daisy	*Leucanthemum vulgare* Lam.	(Native) Scattered in grassland and clearings, particularly in unimproved permanent pasture, but nowhere plentiful.
Wild Privet	*Ligustrum vulgare* L.	(Native; a shrub of base-rich soils) Fairly common in hedges, edges of clearings, wood borders.
Common Toadflax (GCD)	*Linaria vulgaris* Miller	(Native; a common late summer flower of roadside verges and grassy places) In a few places in grassland or beside tracks.
Fairy Flax (GCD)	*Linum catharticum* L.	(Native; a delicate white-flowered *Linum* of dry calcareous grassland) In a few grassy places near woodland, and in unimproved permanent pasture.
Common Twayblade (GCD)	*Listera ovata* (L.) R.Br.	(Native; less conspicuous than other orchids on account of its yellowish-green flowers) In several woodland rides and clearings.
Common Gromwell (GCD)	*Lithospermum officinale* L.	(Native; an uncommon plant of scrubby places, wood borders and hedgerows on basic soils) Recorded in a few places on the edges of woodland; one of the plants to appear at the edge of a plantation after soil disturbance following conifer extraction.
Italian Rye-grass	*Lolium multiflorum* Lam.	(Introduced and widely cultivated) In several places near arable land.
Perennial Rye-grass	*L. perenne* L.	(Native) Fairly common in grassland.
Honeysuckle	*Lonicera periclymenum* L.	(Native) Scattered in hedgerows and open woodland.
Common Bird's-foot Trefoil	*Lotus corniculatus* L.	(Native) Occasional in grassland; less common than expected.
Field Wood-rush	*Luzula campestris* (L.) DC.	(Native; a small, early-flowering rush-like plant of short grassland) Infrequent in woodland grassland and unimproved permanent pasture.
Hairy Wood-rush (GCD)	*L. pilosa* (L.) Willd.	(Native; an ancient woodland species) In three clearings in old woodland.
Ragged-Robin	*Lychnis flos-cuculi* L.	(Native) Occurs sparingly in wet depressions, by pools, the lake, and a stream.
Gipsywort	*Lycopus europaeus* L.	(Native; a waterside plant) Beside the lake, and by one of the pools.
Creeping-Jenny	*Lysimachia nummularia* L.	(Native; in moist grassy places) Fairly common in damp clearings, and beside a stream.

English Name	Latin Name	Notes
Crab Apple	*Malus sylvestris* (L.) Miller	(Native; our only indigenous apple, whose small, sour fruits were gathered by Neolithic man) In several long-established hedges. One old, tall, thorny tree growing in partially open woodland near Field Maple and Wild Service has a girth of 5 ft. 9 ins. (1.75 m) at 5 ft. from ground level.
Musk-mallow	*Malva moschata* L.	(Native; rightly described by the writer Geoffrey Grigson as 'among the prettiest of all English plants') In a few places on the edges of clearings and in open grassland.
Dwarf Mallow	*M. neglecta* Wallr.	(Native) Two plants of this small-flowered *Malva* on a grass verge by a farm track.
Common Mallow	*M. sylvestris* L.	(Native) A few plants near farm buildings.
Pineappleweed	*Matricaria discoidea* DC. (*M. matricarioides* (Less.) Porter nom. illeg.	(Said to have been introduced into Britain in 1871 from Oregon, N. America; now a widespread and common weed of waysides and farmland) Edges of cultivated land, farm tracks and around farm buildings.
Scented Mayweed	*M. recutita* L.	(Native; usually found on light soils) Two plants on arable land on higher ground.
Black Medick	*Medicago lupulina* L.	(Native) Fairly common and widespread in grassland.
Wood Melick	*Melica uniflora* Retz.	(Native; a delicate, pretty woodland grass) On several borders of plantations on ancient woodland sites.
Tall Melilot	*Melilotus altissimus* Thuill.	(Introduced, possibly from Europe in the 16th century. Used in the past by herbalists to make ointments and poultices to reduce swellings, blisters and bruises. Like Woodruff and Sweet Vernal-grass it contains the aromatic oil coumarin, and when dried gives off the delicious scent of new-mown hay) Plentiful in a few clearings, on verges and rides.
Water Mint	*Mentha aquatica* L.	(Native) Frequent by pools, streams, the lake, and in wet ground.
Corn Mint	*M. arvensis* L.	(Native) In several damp, grassy rides and clearings, and on arable land.
Bogbean	*Menyanthes trifoliata* L.	(Native) Around the edges of the lake.
Dog's Mercury	*Mercurialis perennis* L.	(Native; often, but not invariably, an indicator plant of old deciduous woodland) In almost all the open woodland areas; a conspicuous ground-cover plant. At Ditchley it is almost certainly a relic of ancient woodland.

English Name	Latin Name	Notes
Wood Millet	*Milium effusum* L.	(Native; an uncommon, elegant woodland grass) Recorded in open areas of two plantations on ancient woodland sites.
Three-nerved Sandwort	*Moehringia trinervia* (L.) Clairv.	(Native; an inconspicuous woodland herb) At the edges of several open woodland areas.
Field Forget-me-not	*Myosotis arvensis* (L.) Hill	(Native) Common on cultivated land and around farm buildings.
Tufted Forget-me-not	*M. laxa* Lehm. (*M. caespitosa* Schultz)	(Native) A few plants by two of the ponds.
Water Forget-me-not	*M. scorpioides* L.	(Native; one of the larger-flowered forget-me-nots) By ponds and streams.
Spiked Water-milfoil	*Myriophyllum spicatum* L.	(Native) In the deeper water of the lake.
White Water-lily	*Nymphaea alba* L.	(Native) On the lake, in shallow water.
Red Bartsia	*Odontites vernus* (Bellardi) Dumort.	(Native; semi-parasitic on the roots of grasses) Widespread in grassland, conspicuously so in the summer of 1992.
Common Restharrow	*Ononis repens* L.	(Native; a tough, procumbent plant with lovely pink pea-like flowers. Earlier farmers disliked it on ploughed land, because its wiry stems 'arrested the harrow') Rather rare on grassy banks and in old pasture.
Adder's-tongue	*Ophioglossum vulgatum* L.	(Native; a small primitive-looking fern of grassland, open woodland and dune-slacks) Two small groups in woodland grassland, several plants on the edge of a ride, and a few in old pasture. Scarce in west Oxfordshire.
Early-purple Orchid	*Orchis mascula* (L.) L.	(Native) In woodland grassland or rides on old deciduous woodland sites, usually in small numbers, but a fine display of about 300 in a broad ride in 1989.
Wild Marjoram	*Origanum vulgare* L.	(Native; a pretty, aromatic herb whose flowers are attractive to butterflies) Recorded in only two places, in grassland.
Common Broomrape	*Orobanche minor* Smith	(Native; parasitic, mainly on Clover species) One record of two spikes in old grassland.
Wood-sorrel (GCD)	*Oxalis acetosella* L.	(Native) One of the rarer wild flowers of Ditchley. Only occasional flowers seen until 1989, when several fine groups appeared after removal of conifers from a plantation on an ancient woodland site.

English Name	Latin Name	Notes
Prickly Poppy	*Papaver argemone* L.	(Probably native and now rather rare nationally) One plant on edge of oilseed rape field in 1991.
Long-headed Poppy	*P. dubium* L.	(Probably native; less common than *P. rhoeas*) A few plants at the edge of an arable field in 1991.
Yellow-juiced Poppy	*P. dubium* L. subsp. *lecoqii* (Lamotte) Syme	(Native or introduced; sparsely distributed and commoner on the chalk of southern England) A few plants in two different arable fields.
Common Poppy	*P. rhoeas* L.	(Native) Varying numbers each year in standing crops or edges of cultivated fields.
Herb Paris	*Paris quadrifolia* L.	(Native; strongly associated with ancient woodland) Many plants in three different areas of a plantation on an ancient woodland site. One of the most significant discoveries of the survey.
Wild Parsnip	*Pastinaca sativa* L. var. *sylvestris* (Miller) DC.	(Native; an umbelliferous plant of grassland and waysides, especially on chalk and limestone) In several rides and clearings.
Redshank	*Persicaria maculosa* Gray (*Polygonum persicaria* L.)	(Native) A few plants in disturbed ground, and plentiful in damp grassland adjoining an arable field.
Butterbur	*Petasites hybridus* (L.) P.Gaertner, Meyer & Scherb.	(Native) In two separate places by a stream.
Corn Parsley	*Petroselinum segatum* (L.) Koch	(Native; probably rare in west Oxfordshire now) On a few widely separated grass verges bordering arable fields. A pleasing addition to the Ditchley list.
Reed Canary-grass	*Phalaris arundinacea* L.	(Native; a tall reed-like grass of wet habitats) Beside the lake and by one of the pools.
Small Cat's-tail (GCD)	*Phleum bertolonii* DC.	(Native; similar in appearance to Timothy, but much smaller) Infrequent, mainly in old grassland and unimproved pasture.
Timothy	*P. pratense* L.	(Native) Here and there in grassland.
Hart's-tongue (GCD)	*Phyllitis scolopendrium* (L.) Newman	(Native) Recorded in only one place, in stonework, in 1988, but by 1990, after two dry summers, the plant died.
Norway Spruce (P)	*Picea abies* (L.) Karsten	(Introduced) Widely used in the plantations as a nurse tree for beech.

English Name	Latin Name	Notes
Bristly Oxtongue	*Picris echioides* L.	(Probably introduced; well naturalized in waste places and on rough ground) A few plants on disturbed ground in 1988.
Mouse-ear Hawkweed (GCD)	*Pilosella officinarum* F. Schultz & Schultz-Bip. (*Hieracium pilosella* L.)	(Native; a charming, small hawkweed bearing lemon-yellow flowers) In a few places in old grassland and unimproved permanent pasture.
Burnet-saxifrage	*Pimpinella saxifraga* L.	(Native; a delicate umbellifer of dry calcareous grassland) In unimproved permanent pasture and on grassy banks. Infrequent.
Corsican Pine (P)	*Pinus nigra* subsp. *laricio* Maire	(Introduced) Occasionally used as a nurse tree for beech in the plantations.
Scots Pine (P)	*P. sylvestris* L.	(Native) Occasionally used as a nurse tree for beech in the plantations. Several fine mature trees left to disperse seed.
Ribwort Plantain	*Plantago lanceolata* L.	(Native) Common and widespread.
Greater Plantain	*P. major* L.	(Native) Common and widespread, usually on well-trodden ground.
Hoary Plantain	*P. media* L.	(Native; less common than the preceding two species and usually on basic soils. Very attractive when in flower, the stamens being a lilac-pink colour) In a few places in grassy rides, tracks and unimproved pasture.
Greater Butterfly Orchid (GCD)	*Platanthera chlorantha* (Custer) Reichb.	(Native) One of the notable wild flowers of Ditchley, now rare in west Oxfordshire. In a few places on the borders of woodland or in grassy rides, not always flowering.
London Plane (P)	*Platanus* × *hispanica* Miller ex Muenchh. (*P. occidentalis* L. × *P. orientalis* L.)	(Introduced) One in narrow belt of amenity woodland.
Narrow-leaved Meadow-grass	*Poa angustifolia* L.	(Native) Recorded in an area of woodland grassland.
Annual Meadow-grass	*P. annua* L.	(Native) Common and widespread in grass, beside tracks, on bare ground.
Wood Meadow-grass	*P. nemoralis* L.	(Native) Woodland borders.
Smooth Meadow-grass	*P. pratensis* L.	(Native) Common in grassland.
Rough Meadow-grass	*P. trivialis* L.	(Native) Common in grassland.

English Name	Latin Name	Notes
Common Milkwort (GCD)	*Polygala vulgaris* L.	(Native) Recorded rarely, in old, short grassland.
Equal-leaved Knotgrass	*Polygonum arenastrum* Boreau	(Native; less common than the following species) A few plants at edge of flax (linseed) field in 1991.
Knotgrass	*P. aviculare* L.	(Native) Common around the edges of cultivated fields, farmyards, bare ground.
Polypody	*Polypodium vulgare* L.	(Native) Several groups of this attractive fern in two widely separated stone walls near buildings.
Poplar hybrids (P)	*Populus* 'Robusta', 'Serotina' and TT32	(Introduced) Planted by streams and on damp, low-lying land in several places on the estate.
Small Pondweed	*Potamogeton berchtoldii* Fieber	(Native; a fine-leaved pondweed susceptible to pollution) In deep water in the lake.
Curled Pondweed	*P. crispus* L.	(Native; common in lowland lakes, ponds, etc.; very nitrogen-resistant) In deep water in the lake.
Silverweed	*Potentilla anserina* L.	(Native) Fairly common on grassy tracks, trampled ground, edges of clearings.
Tormentil	*P. erecta* (L.) Raeusch	(Native; more common on acid soils) In a few places in old, short grassland.
Creeping Cinquefoil	*P. reptans* L.	(Native) Widespread and fairly common in open grassland.
Barren Strawberry	*P. sterilis* (L.) Garcke	(Native; sometimes mistaken for wild strawberry) In all the woodland margin areas and most of the clearings.
Cowslip	*Primula veris* L.	(Native) One of the most valued wild flowers at Ditchley. Luxuriant in unimproved permanent pasture, and in varying numbers in most of the rides and remnants of old grassland.
Primrose (GCD)	*P. vulgaris* Hudson	(Native; now scarce in west Oxfordshire) Plentiful and widespread in some rides, or in remnant deciduous woodland
False Oxlip	*P. vulgaris* × *P. veris* (*P.* × *polyantha* Miller)	(Native; a natural hybrid between primrose and cowslip) Recorded in numerous places where the two species grow near each other; many specimens in some years.
Selfheal	*Prunella vulgaris* L.	(Native) Common and widespread in grassland.

English Name	Latin Name	Notes
Wild Cherry (P)	*Prunus avium* (L.) L.	(Native) Occurs in a few hedges. Widely planted, often along the edges of plantations, where it is a lovely sight in spring.
Blackthorn	*P. spinosa* L.	(Native) A common shrub, frequent in hedges, on edges of woodland, and in small scrubby corners bordering grassland.
Douglas Fir (P)	*Pseudotsuga menziesii* (Mirbel) Franco	(Introduced) Widely used in the plantations as a nurse tree for beech.
Bracken	*Pteridium aquilinum* (L.) Kuhn	(Native) Common and widespread, in spite of its well-known preference for acid soils. Dominant in many of the rides, to the detriment of smaller and more sensitive species, and often hiding the autumn-flowering Meadow Saffron.
Common Fleabane	*Pulicaria dysenterica* (L.) Bernh.	(Native; a plant of wet meadows, marshes, watersides and ditches, bearing golden-yellow daisy-like flowers) Beside the lake and by one of the ponds.
Wild Pear	*Pyrus pyraster* (L.) Burgsd.	(Introduced; many are probably stocks of cultivated trees) One tall old tree in a boundary hedge.
Pedunculate Oak	*Quercus robur* L.	(Native) Many fine old and young trees in the parkland, remnants of deciduous woodland, and in hedgerows.
Meadow Buttercup (GCD)	*Ranunculus acris* L.	(Native) Seldom recorded in old pasture.
Goldilocks Buttercup	*R. auricomus* L.	(Native; open woodland and hedgebanks; an ancient woodland species) Surprisingly rare. Recorded only twice, on a grassy bank and beside a track.
Bulbous Buttercup	*R. bulbosus* L.	(Native; an early-flowering buttercup of dry grassland) In several localities in old pasture and dry clearings.
Lesser Celandine	*R. ficaria* L.	(Native) Frequent in moist places in open woods, hedgebanks and on verges.
Greater Spearwort	*R. lingua* L.	(Native; one of the rarer species of *Ranunculus*) A small group in shallow water in the lake.
Creeping Buttercup	*R. repens* L.	(Native) Common and widespread in all damp grassland areas.

English Name	Latin Name	Notes
Thread-leaved Water-crowfoot	*R. trichophyllus* Chaix	(Native; one of the white-flowered crowfoots of ponds, ditches, streams and rivers) One record only: a small group of plants in a shallow muddy pool below a spring-line. (It is interesting that this is the only water-crowfoot recorded in Wychwood, in two of the ponds.)
Wild Radish	*Raphanus raphanistrum* L.	(Probably introduced and said to have been present in prehistoric times) A few plants in vehicle ruts in a woodland ride.
Wild Mignonette	*Reseda lutea* L.	(Native; usually on calcareous soils) On a few grassy tracks or edges of arable land.
Weld	*R. luteola* L.	(Native; described by the writer Geoffrey Grigson as '. . . one of the most ancient and one of the best of the dyer's plants . . . giving specially brilliant, pure, and fast yellows.') Rarely recorded, but several plants appeared in disturbed ground at the edge of a ride after conifer extraction.
Buckthorn	*Rhamnus cathartica* L.	(Native; the leaves are the foodplant of the caterpillar of the Brimstone butterfly) Uncommon. Noted in only a few hedges or borders of old woodland.
Yellow-rattle	*Rhinanthus minor* L.	(Native) Frequent in a number of woodland rides and clearings.
Red Currant	*Ribes rubrum* L.	(Probably introduced; in hedges, woods and scrub, often an escape) Recorded in three separate places on woodland borders.
Gooseberry	*R. uva-crispa* L.	(Probably native) In a few places on woodland edges; possibly bird-sown.
False-acacia (P)	*Robinia pseudoacacia* L.	(Introduced; native of N. America) Several mature trees by an estate road, a few self-sown saplings on a rough verge, and several tall stems which have developed from old stumps in a double hedge.
Water-cress	*Rorippa nasturtium-aquaticum* (L.) Hayek	(Native) In several places by ponds and beside the lake.
Field-rose	*Rosa arvensis* Hudson	(Native; a trailing white-flowered rose sometimes forming dense mounds on the borders of hedgerows and woodland) Fairly common, a few of these tangled bushes being of great size and lovely when in flower.
Dog-rose	*R. canina* L.	(Native) Widespread in hedgerows, and occasionally as free-standing bushes.
Harsh Downy-rose	*R. tomentosa* Smith	(Native; uncommon in west Oxfordshire) Two hedgerow bushes, one of which was unfortunately destroyed during conifer extraction.

English Name	Latin Name	Notes

The following nine *Rosa* hybrids and *Rosa* groups were identified by The Rev. A.L. Primavesi (*Rosa* specialist, Botanical Society of the British Isles), from specimens (fruits, leaves, and thorn-bearing stems) sent to him in the autumn of 1990 and 1991.

Field-rose × Dog-rose	*R. arvensis* × *R. canina* (*R.* × *verticillacantha* Mérat)	(Native) In a small patch of scrub.
Hairy Dog-rose × Dog-rose	*R. caesia* subsp. *caesia* × *R. canina*	(Native) In three different hedgerows.
Glaucous Dog-rose × Dog-rose	*R. caesia* subsp. *glauca* × *R. canina*	(Native) Two bushes in a hedgerow and another on the edge of a clearing.
Dog-rose × Harsh Downy-rose	*R. canina* × *R. tomentosa* (*R.* × *scabriuscula* Smith)	(Native) Hedge on south side of a plantation.
Harsh Downy-rose × Dog-rose	*R. tomentosa* × *R. canina*	(Native) In two places at the edges of rides.
Dog-rose (*R. canina*)	Group *Dumales*	(Native) On the edge of a plantation.
Dog-rose (*R. canina*)	Group *Lutetianae*	(Native) In a small piece of scrub and in a hedgerow.
Dog-rose (*R. canina*)	Group *Pubescentes*	(Native) Three bushes in hedgerows and one on an open grassy bank.
Dog-rose (*R. canina*)	Group *Transitoriae*	(Native) Three bushes in a hedge adjoining a ride.
Dewberry	*Rubus caesius* L.	(Native) Occasionally recorded at edges of woodland.
Bramble	*R. fruticosa* L. agg.	(Native) Common in hedgerows, edges of rides and clearings, or on the woodland floor.
Raspberry	*R. idaeus* L.	(Native) A few canes, bearing fruit, on the edge of old woodland.
Common Sorrel	*Rumex acetosa* L.	(Native) Occasional in old grassland.
Clustered Dock	*R. conglomeratus* Murray	(Native; a dock of damp habitats) Beside the lake and by one of the ponds.
Curled Dock	*R. crispus* L.	(Native) In numerous places in grassland.
Broad-leaved Dock	*R. obtusifolius* L.	(Native; the commonest dock) Widespread in grassland, beside tracks, edges of farmland.
Wood Dock (GCD)	*R. sanguineus* L. var. *viridis* (Sibth.) Koch	(Native; a dock of woodland borders and damp, shady hedgerows) Infrequent in habitats of this kind, but possibly overlooked.

English Name	Latin Name	Notes
Procumbent Pearlwort	*Sagina procumbens* L.	(Native) Recorded once, in a vehicle rut in one of the rides.
Goat Willow	*Salix caprea* L.	(Native) A small number of young and mature trees in hedgerows, beside ponds and the lake.
Grey Willow	*S. cinerea* L. subsp. *oleifolia* Macreight	(Native) Commoner than the above species; near ponds, in wet scrub, damp corners of hedges, and beside the lake.
Crack Willow	*S. fragilis* L.	(Native) Several old trees in a marshy area in the bottom of a meadow, and a few by ponds and the lake. An uncommon tree at Ditchley.
Meadow Clary or Meadow Sage	*Salvia pratensis* L.	(Native; a beautiful plant, very scarce nationally) Several hundred plants in an old meadow, where they have existed for centuries. An important site, both locally and nationally.
Elder	*Sambucus nigra* L.	(Native) Common and widespread in hedges, and on margins of plantations and old woodland.
Salad Burnet (GCD)	*Sanguisorba minor* Scop. subsp. *minor*	(Native; a calcareous grassland herb) Infrequently distributed in old unimproved pasture, clearings and rides.
Great Burnet	*S. officinalis* L.	(Native; a locally scarce species of old wet meadows) Several small groups of this attractive plant on a bank above a stream.
Sanicle	*Sanicula europaea* L.	(Native; an ancient woodland plant) One of the rarer plants, recorded in only a few places, in small numbers, on edges of plantations on ancient woodland sites.
Common Club-rush	*Schoenoplectus lacustris* (L.) Palla	(Native) In one of the small ponds, and in shallow water in the lake.
Water Figwort	*Scrophularia auriculata* L.	(Native) By streams and ponds, and beside the lake.
Common Figwort	*S. nodosa* L.	(Native) Occasionally recorded in rides, clearings and woodland borders.
Hoary Ragwort	*Senecio erucifolius* L.	(Native; an attractive ragwort, not to be confused with the Common Ragwort of poor pastures) In several clearings and rides.
Common Ragwort	*S. jacobaea* L.	(Native) Uncommon, and always in small numbers, in grassland.

English Name	Latin Name	Notes
Oxford Ragwort	*S. squalidus* L.	(Introduced; believed to have escaped from Oxford Botanic Garden, first recorded on walls in the city in 1794 and now widespread) A few plants on disturbed ground.
Groundsel	*S. vulgaris* L.	(Native) Here and there on cultivated or disturbed ground.
Field Madder	*Sherardia arvensis* L.	(Native) Scarce, and in limited numbers around the edges of cultivated land.
Pepper-saxifrage	*Silaum silaus* (L.) Schinz & Thell.	(Native; locally scarce) A few plants in damp grass beside a stream.
Red Campion	*Silene dioica* (L.) Clairv.	(Native) Of rare occurrence on the edges of remnant deciduous woodland.
White Campion	*S. latifolia* Poiret	(Native, or introduced in Neolithic times) The commonest of the four species of *Silene* recorded. Sparsely distributed by hedgerows, on cultivated land and at the edges of clearings.
Night-flowering Catchfly (GCD)	*S. noctiflora* L.	(Native. A very scarce cornfield flower; night-flying insects are attracted to the sweet, clover-like scent of the flowers) Another of the 25 arable species selected for survey by the B.S.B.I. in 1986/87. Recorded, in sparse numbers, every year in one or other of the cultivated fields, where it is hoped it will persist.
Bladder Campion	*S. vulgaris* Garcke	(Native) In small numbers in a variety of grassy habitats.
Charlock	*Sinapis arvensis* L.	(Probably native) Rarely recorded, on disturbed ground or in crops.
Hedge Mustard	*Sisymbrium officinale* (L.) Scop.	(Native) Infrequent, around cultivated fields and farm buildings.
Bittersweet	*Solanum dulcamara* L.	(Native) Occurs occasionally in hedges or near ponds and streams.
Canadian Goldenrod	*Solidago canadensis* L.	(Introduced; commonly grown in gardens) One group of stems on the edge of a poplar plantation; probably a garden escape.
Perennial Sow-thistle	*Sonchus arvensis* L.	(Native) Recorded in two places, on farmland and waste ground.
Prickly Sow-thistle	*S. asper* (L.) Hill	(Native) Infrequent on disturbed ground, verges, cultivated land.
Smooth Sow-thistle	*S. oleraceus* L.	(Native) Occasional near farm buildings.

English Name	Latin Name	Notes
Rowan	*Sorbus aucuparia* L.	(Native) Recorded in only two places, where it was probably planted.
Wild Service-tree (GCD)	*S. torminalis* (L.) Crantz	(Native; strongly associated with ancient woodland) Some fine young and old trees in remnant deciduous woodland, hedgerows, and old parkland. (See main text for details.)
Field Woundwort	*Stachys arvensis* (L.) L.	(Native; normally an arable species of non-calcareous soils) Numerous plants found in late summer 1989 in two fields from which oilseed rape and broad beans had been harvested.
[Downy Woundwort	*S. germanica* L.	(Native. Very rare and endangered or vulnerable; only four sites in Great Britain in 1992. Confidential records held by English Nature.) One site in the vicinity of Ditchley.]
Betony (GCD)	*S. officinalis* (L.) Trev. St. Léon	(Native; still used by medical herbalists to relieve headaches, neuralgia, cuts and bruises) Many large and small groups of this attractive plant in clearings and rides, and on open woodland borders.
Hedge Woundwort	*S. sylvatica* L.	(Native) Common and widespread in hedgerows, clearings and on wood margins.
Lesser Stitchwort	*Stellaria graminea* L.	(Native; usually on light soils) In one of the clearings, and in two areas of damp grassland.
Greater Stitchwort	*S. holostea* L.	(Native) Plentiful in a section of old deciduous woodland after the felling of hardwoods (mainly Ash) in 1987; uncommon elsewhere, by hedgerows and open wood borders.
Common Chickweed	*S. media* (L.) Villars	(Native) Sparsely distributed on cultivated land and disturbed ground.
Bog Stitchwort	*S. uliginosa* Murray	(Native) By one of the streams.
Devil's-bit Scabious	*Succisa pratensis* Moench	(Native; now uncommon in west Oxfordshire) Recorded in only two places: in a small area of woodland grassland, and in old pasture near a stream.
Snowberry (P)	*Symphoricarpos albus* (L.) S.F. Blake (*S. rivularis* Suksd.)	(Introduced; native of N. America) Occasionally planted as game cover.
Russian Comfrey	*Symphytum × uplandicum* Nyman	(Originally introduced as fodder; widely cultivated on the Continent as cattle food and now the commonest comfrey) One record only, on the edge of a meadow.

English Name	Latin Name	Notes
Black Bryony	*Tamus communis* L.	(Native; a climbing plant, trailing over hedges, leaving skeins of yellow and red berries in autumn) Infrequent in hedgerows.
Common Dandelion	*Taraxacum officinale* Wigg. group (Section *Ruderalia*)	(Native) Widespread and occasionally plentiful.
(No English name)	*T. oxoniense* Dahlst. (Section *Erythrosperma*)	(Native; a microspecies of *Taraxacum* with sharply dissected narrow-lobed leaves, small heads and red, purple or violet-coloured seeds) Numerous plants in unimproved permanent pasture.
Yew (GCD)	*Taxus baccata* L.	(Native) A few mature trees, and many young ones introduced into the plantations.
Field Pennycress	*Thlaspi arvense* L.	(Possibly native) Occasional plants at the margins of a few arable fields.
Lime (P)	*Tilia × vulgaris* Hayne	(A hybrid between Large-leaved Lime (*Tilia platyphyllos*) and Small-leaved Lime (*T. cordata*). Rarely native, but widely planted) Many fine mature trees in the parkland and by estate roads, and also newly-planted avenues.
Upright Hedge-parsley	*Torilis japonica* (Houtt.) DC.	(Native; a smaller, neater relative of Cow Parsley, flowering later) Occurs on the edges of many clearings and hedgebanks.
Goat's-beard	*Tragopogon pratensis* L.	(Native) In a few places in clearings and rides.
Hop Trefoil	*Trifolium campestre* Schreber	(Native; so called because when the flower-heads are dry they resemble miniature hops) Uncommon in a few rides and clearings.
Lesser Trefoil	*T. dubium* Sibth.	(Native) Infrequent in drier grassy places.
Zigzag Clover	*T. medium* L.	(Native; somewhat local in distribution) Rare, having been recorded in only one section of grassland in a woodland clearing.
Red Clover	*T. pratense* L.	(Native) Scattered in grassy areas.
White Clover	*T. repens* L.	(Native) Often occurs with Red Clover, in similar places.
Scentless Mayweed	*Tripleurospermum inodorum* (L.) Schultz-Bip.	(Native; still a fairly common arable plant) Occurs occasionally at edges of some of the arable fields, or in crops.
Yellow Oat-grass (GCD)	*Trisetum flavescens* (L.) P. Beauv.	(Native; usually on calcareous soils) In several remnants of old grassland and in unimproved permanent pasture.

English Name	Latin Name	Notes
Colt's-foot	*Tussilago farfara* L.	(Native) Small groups of plants in a number of rides and clearings, or on rough ground.
Bulrush	*Typha latifolia* L.	(Native) In shallow water in the lake.
Wych Elm (P)	*Ulmus glabra* Hudson	(Native) Numerous suckers in one of the smaller plantations and in a hedgerow beside a stream; introduced into a few of the plantations.
English Elm (P)	*U. procera* Salisb.	(Probably native; rarely seen as a mature tree since the destruction of elms by Dutch Elm Disease in the 1970s) Among many suckers noted in hedgerows a few have attained the size of small trees, which at present appear healthy. One of the hardwood trees introduced into several of the plantations.
Common Nettle	*Urtica dioica* L.	(Native) Common and widespread.
Small Nettle	*U. urens* L.	(Probably native) In Ellen's Lodge garden, and in an adjoining piece of ground previously used as a chicken-run.
Marsh Valerian	*Valeriana dioica* L.	(Native; a locally scarce plant of marshy ground) In two locations, beside streams.
Common Valerian (GCD)	*V. officinalis* L.	(Native; one of the most valuable of the medical herbalist's plants, the constituent properties of the roots alleviating nervous tension, insomnia and headaches) In almost all the damp rides, wet hollows and margins of water.
Narrow-fruited Cornsalad (GCD)	*Valerianella dentata* (L.) Pollich	(Native; an increasingly scarce, inoffensive cornfield flower) In one place only, at the edge of a wheat field, in 1990.
Dark Mullein	*Verbascum nigrum* L.	(Native; mainly on limestone soils) Several plants in the Timberyard area.
Great Mullein	*V. thapsus* L.	(Native; mostly on sandy or calcareous soils) After soil disturbance following conifer extraction in one of the plantations, two flowering plants and 83 non-flowering rosettes appeared on the adjoining ride. A few plants have flowered in the same place since then.
Green Field-speedwell	*Veronica agrestis* L.	(Native; an uncommon speedwell of cultivated ground) Recorded in two such places.
Blue Water-speedwell	*V. anagallis-aquatica* L.	(Native) Beside the lake, and in quantity in one of the pools.

English Name	Latin Name	Notes
Wall Speedwell	*V. arvensis* L.	(Native) In a few places, in bare ground.
Brooklime	*V. beccabunga* L.	(Native) Present in almost all wet places – by ponds, streams, the lake, and in depressions and hollows where water lies.
Germander Speedwell	*V. chamaedrys* L.	(Native; one of the prettiest speedwells) Widely dispersed on woodland margins, edges of fields and hedges, and the shadier parts of clearings.
Slender Speedwell	*V. filiformis* Smith	(Introduced as a rock garden plant in the early 1900s and now widespread; often in lawns. Native of the Caucasus) In a few places on closely-cut grass verges or parkland.
Ivy-leaved Speedwell	*V hederifolia* L.	(Native; usually a plant of cultivated land, and sometimes a persistent garden weed) A small group of plants outside the boundary fence of one of the plantations.
Wood Speedwell	*V. montana* L.	(Native) A few plants on a damp bank by one of the pools. Surprisingly rare at Ditchley. It is the commonest speedwell in Wychwood, where many of the rides are wetter and the woods damper.
Heath Speedwell (GCD)	*V. officinalis* L.	(Native; prefers well-drained acid soils) Noted in a few places in short dry grass or bare patches of soil in open situations.
Common Field Speedwell	*V. persica* Poiret	(Introduced. First recorded in 1825 and now common on cultivated land throughout Britain. Native of SW Asia) One of the commonest plants to be found on the edges of arable fields and bare tracks.
Thyme-leaved Speedwell	*V. serpyllifolia* L.	(Native) In numerous places in grassland, and in unimproved permanent pasture.
Wayfaring-tree	*Viburnum lantana* L.	(Native; mostly on base-rich soils) Fairly common in hedges and on margins of open woodland.
Guelder-rose	*V. opulus* L.	(Native) Occasional on the edges of plantations and old woodland, or on scrubby banks, the rich colours of the foliage and fruits enhancing these places in autumn.
Tufted Vetch	*Vicia cracca* L.	(Native) Uncommon by hedgerows or at the edges of a few clearings.
Hairy Tare	*V. hirsuta* (L.) Gray	(Native) In a few places in long grass in open situations.

English Name	Latin Name	Notes
Common Vetch	*V. sativa* subsp. *nigra* (L.) Ehrh.	(Native or introduced) In several places in grassland, but always in limited numbers.
Bush Vetch (GCD)	*V. sepium* L.	(Native) Widespread and fairly common in hedgerows, on edges of clearings and verges, and occasionally in open woodland.
Wood Vetch	*V. sylvatica* L.	(Native; rarely recorded in west Oxfordshire, and does not occur in Wychwood) A spectacular display on the edge of a plantation after the felling of conifers in 1988; subsequently found in small quantities on a few wood borders or rough rides. Another of Ditchley's special plants.
Smooth Tare (GCD)	*V. tetrasperma* (L.) Schreber	(Native) Less common than Hairy Tare, but growing in similar places.
Greater Periwinkle	*Vinca major* L.	(Introduced) One patch on a verge beside a track leading to one of the farms; probably a garden 'throw-out'.
Field Pansy	*Viola arvensis* Murray	(Native) Fairly frequent on disturbed ground or at the margins of arable fields.
Hairy Violet (GCD)	*V. hirta* L.	(Native; a beautiful violet, not of woodland but of calcareous grassland, often growing in large clumps) In considerable numbers on a rough, steep grassy bank; occasional elsewhere in clearings, rides, and old unimproved pasture.
Sweet Violet	*V. odorata* L.	(Native) Plants bearing violet or white flowers in many warm, sheltered corners of old woodland, and on hedgebanks.
Early Dog-violet (GCD)	*V. reichenbachiana* Jordan ex Boreau	(Native; an early-flowering, paler violet) Not uncommon in light shade on the edges of old woodland and occasionally in shady hedgebanks.
Common Dog-violet	*V. riviniana* Reichb.	(Native) Occurs in almost all deciduous woodland areas and at the edges of plantations on the ancient woodland sites, usually with, or near, the above two species.

Wild Service Tree

Sorbus torminalis L.